COLLECTING AND RESTORING ANTIQUE FIRE ENGINES

BY ROBERT LICHTY

MODERN AUTOMOTIVE SERIES

TAB BOOKS Inc.
BLUE RIDGE SUMMIT, PA. 17214

FIRST EDITION

FIRST PRINTING

JANUARY 1981

Printed in the United States of America

Library of Congress Cataloging in Publication Data

Lichty, Robert.
Collecting and restoring antique fire engines.

Includes index.
1. Fire-engines—Collectors and collecting.
I. Title.
TH9371.L52 628.9'252 80-20666
ISBN 0-8306-9700-4
ISBN 0-8306-2099-0 (pbk.)

Contents

Acknowledgements

Special thanks are due to the following people who in one way or another contributed directly to the completion of this book. Many through support and information, others through more direct contributions; but to all of them, thank you.

Donald Adams of Greenfield Village, Dearborn, Michigan; John Bergquist of The Antique Fire House, Loveland, Colorado; Mike Boucher of Firetrucks Unlimited, Englewood, Colorado; Terry Boyce, Hutchinson, Kansas; Anthony Butera and Michael Carbonella, Staten Island, New York; R.C. Darley of W.S. Darley Co., Melrose Park, Illinois; John "Gunner" Gunnell, R. Chris Halla and Tony Hossain, all of *Old Cars Weekly*, Iola, Wisconsin; Chester L. Krause, Iola, Wisconsin; John Kress, Jr., Sparta, Wisconsin; Doug Ogilvie, president of Pierce Manufacturing, Appleton, Wisconsin; Ken Soderbeck of Jackson, Michigan; Trucks Unlimited; Car Exchange Magazine; Waterous Company; Henry Ford Museum, Dearborn, Michigan; Hall of Flame, Phoenix, Arizona; and Bob Milnes.

A special note of thanks to Bob Lemke of Iola, Wisconsin. This book would not have reached its level of quality without his help in editing this manuscript and through direct contributions.

And a double special thanks to my wife Donna who has not only tolerated my sickness for old vehicles but shared my enthusiasm in antique fire engines. And to my son Christopher who patiently waited at home while daddy hammered away about fire engines at the office. Plus an advance thank you to his brother or sister waiting to be born at the time of this writing and who will have to learn from Chris and his mother the patience that having a fire buff for a father will require.

Preface

Probably no other type of collector vehicle offers the hobbyist more fun per dollar than a shiny restored antique fire engine. Bright paint, brass, copper, chrome, nickel, bells, siren, lights, hoses, ladders and pumps, topped off by gold leaf pin striping and lettering add up to one spectacular vehicle that attracts attention wherever it goes. Children from one to one hundred have always been fascinated by firemen and their giant machines that roar through the streets on life-saving missions.

Best of all, in today's collector vehicle marketplace the fire engine is usually a bargain compared to an antique auto of the same vintage and state of preservation. Add the exclusivity of highly limited production, plus a wealth of custom equipment and the appeal of a big gleaming fire truck is difficult to resist.

However, a basically low purchase price is only the beginning. The collector must find a vehicle suited to the limitations of his own mechanical skills as well as his restoration project budget. Then comes research of the truck manufacturer and the body-maker who originally put the truck on the road.

Even more important, (and more fun) is the research on the history of the truck itself. It is especially necessary for an authentic restoration. A decision must be made on how far to go with the restoration and to what uses the fire engine will be put when the project is completed. Will it be a national show winner? A parade truck? A participant in antique fire equipment musters? Or just a static display piece in a museum or in front of a theme restaurant? Budget and ambition will dictate whether the collector/restorer chooses a rough vehicle with a great deal of work to be done, or a clean original truck, or even an already-restored rig.

But where do you garage and work on a 55-foot aerial ladder truck once you own it? Where and how do you start the monumental task of a complete restoration? Short of a muscle-building program, how do you wrestle 300-pound parts around your garage? And where do you get the parts you need?

What about when the masterpiece is completed? There are parades, shows and musters to attend. How do you find out where such events are held? What clubs do you join to meet other antique fire equipment enthusiasts? Where are the fire museums in the country that can provide inspiration and information? And what types of related fire memorabilia are the most collectible?

We will answer all of these questions and more to make the prospect of a fire engine restoration less frightening and to put a better perspective on the rewards. We won't be going into detail in the areas already well covered by TAB BOOKS Inc. antique auto restoration books and those of other publishers. We still, however, cover the major differences between fire apparatus restoration and similar work more commonly encountered in the restoration of antique autos. And there will be special emphasis on the unique challenges and rewards found in no other type of historical vehicle refurbishing project. Up-to-date listings of clubs, manufacturers and suppliers in the appendices should also prove useful as the restoration work proceeds.

It is hoped that the project you decide to tackle brings as much fun and personal gratification to you and your ever-patient family as we have personally been able to experience in the fascinating world of antique fire engines.

Robert Lichty

1

Fire Engines as Collectibles

Since the days of the horse-drawn steam fire pumper, little children have been fascinated by the monstrous machines that clattered and clanged through their neighborhood streets on missions of rescue (Fig. 1-1).

CHILDHOOD FANTASIES

Those little children have grown up, but for many of them the fascination for fire equipment lives on. Some became firemen and now live out the fantasies those machines evoked. Others, while their lives may have taken a less exciting turn, nurtured their early interest and have become what are known as fire buffs: historians of fires, firemen and the equipment used to battle that most terrifying of natural and man-made disasters—fire.

Many fire buffs specialize in the area of fire-fighting apparatus, especially the wide range of fire engines. The vintage of such fire equipment that attracts the fire buff is often dependent on the collector's age and what type of fire engine brings back the most vivid childhood memories. This pull of nostalgia plays one of the most significant roles in determining why a person decides to collect almost anything; especially anything as impractical as a giant red truck loaded with complex accessories. A fire buff in his fifties may re-

Fig. 1-1. It was horse-drawn steam pumpers that invoked fantasies of being a fireman in the hearts of many children at the turn of the century. This 1906 Ahrens steamer is a classic example.

member most fondly the American La France, with its solid white rubber tires, that was garaged at the fire house just down the block; or, perhaps the Model T Ford fire engine used by the rural volunteer fire department. The 20- or 30-year old, on the other hand, is quite likely to have a strong attraction for the Series 700 American La France trucks built in the late 1940s and early 1950s. The city boy may remember a huge fleet of complex and fancy piston-pumping Ahrens Fox fire engines scattered among fire companies all over town, while the country boy may have the fondest recollections of the single local truck that was built on a Ford, Chevrolet or Studebaker chassis and on which his father may have served as a volunteer smoke-eater.

HOBBY COLLECTING

As a collector becomes more sophisticated, or just plain addicted to the hobby of fire engine collecting and restoration, he may advance to the more elaborate trucks out of fascination for the complexity or capabilities of the unit (Fig. 1-2). It is not uncommon to find a fire buff living in a midwestern town of a few hundred people who owns a 75- or 100-foot aerial ladder truck— even though the tallest building in his town may only be two stories. As a fire buff's involvement with the hobby grows, he is likely to become more active in clubs and group events called *musters*. Details on the musters will be covered in another chapter, but they are often directly responsible for the addition of more equipment to a collection: The more competitive the hobbyist becomes in such events, the more desire he has to increase his equipment collection.

And then there is the "I bought it because it was there" syndrome. This is possibly the most dangerous sign of fire engine addiction because a collector will hear of a vehicle being put up for sale by a local fire department, submit a bid and, lo and behold, he owns another fire truck (Fig. 1-3). Of course, the less practical the unit is, the more likely the fire buff is to get it. The collector seldom holds the winning bid on the nice little 20-foot pumper or squad unit, but somehow

Fig. 1-2. What manner of fire buff would pay the enormous expense of importing an English Dennis aerial ladder truck across the Atlantic? In this case, John Kress, a typical fire engine collector and department store owner from Sparta, Wisconsin.

Fig. 1-3. Why would anyone want a 1926 Dodge chemical truck? One may answer that question with visions of many fun-filled hours in parades, musters and shows. The Edgewater, Colorado Fire Department can tell you why.

always manages to win the aforementioned 100-foot aerial ladder rig. The plight of finding suitable storage facilities for a 50-foot long fire engine can only be surpassed by trying for the first time to tune a V-12 engine with twin ignition and 24 spark plugs.

However, that's what makes the hobby of collecting fire apparatus different from collecting just an old car or truck. The nostalgia, the glamour, the engineering and the sheer mass of fire engines make them irrestible to many people. Chances are that since you are reading this book, you are one such person. If there is not a giant red fire fighter in your driveway yet, there soon may be.

2

Types, Sizes and Uses of Apparatus

Before the would-be fire engine collector buys that all-important first fire engine, he should have some idea of what he wants and what is available. Knowing what the different types of fire trucks are and what they were meant to do will help insure that the initial fire engine purchase is the right first step in becoming involved in this great hobby.

PUMPERS

The fire pumper (Fig. 2-1) is usually a fairly small truck, as compared in length to the ladder trucks. It is used to draw water from the hydrant or other water source to the fire while developing enough pressure to deliver the water to where the firemen need it to fight the blaze. The pumps that perform that work can vary from the rotary gear-driven type to the piston type for which Ahrens Fox equipment was famous for so many years. Most pumpers carry a sizable complement of hoses, nozzles and other related equipment. Usually the pumper will carry on its side or over a front fender a large heavy hose, called a *suction hose.* It is used when water must be drawn from a source that is not under pressure, such as a pond or a portable water tank. The softer light-colored hose carried on the pumper is used for connecting to hydrants and for the

Fig. 2-1. The fire pumper is probably one of the most conveniently sized fire engines for the collector/restorer. Not as long as the ladder trucks, it is more likely to fit in the average garage. This 1944 Dodge-Darley commercial chassis pumper is shown at the factory before fully equipped with hose, etc.

actual fire battle. It is usually carried inside the truck and folded in the rear or on a reel.

CHEMICAL TRUCKS

The chemical truck (Fig. 2-2) is similar to the pumper in overall size. Its main function is to carry chemical-type fire fighting equipment to the scene of a fire. The large tanks on the chemical truck work like giant fire extinguishers. Reflecting the state of the art of chemical fire fighting methods, older chemical trucks have small capacities; but, as the units have become more modern in response to the growing technology in the field, they have become bigger and now carry great quantities of fire extinguishing materials. Probably the most familiar type of chemical truck is the airport fire engine which stands by to battle the explosive blazes generated by modern high octane aviation fuel.

TANKERS

The tanker (Fig. 2-3) carries water to a fire. Tankers are more prevalent in communities and rural areas which do not have pressurized hydrants to supply water to the entire terri-

Fig. 2-2. Although chemical trucks may not offer as much muster fun, this 1929 Reo/Boyer proves they can be spectacular in full trim.

tory covered by a fire district. It is not uncommon in many parts of the country for pumpers and tankers alike to be equipped with four-wheel drive, allowing them access to rough terrain, as in the case of a forest fire.

LADDER TRUCKS

The ladder truck (Fig. 2-4) can be a straight chassis unit carrying anything thing from an assortment of ground ladders to the giant 75-and 100-foot aerial ladder trucks (sometimes referred to as *hook-and-ladder trucks*). By virtue of their sheer size, ladder trucks tend to be the least practical fire engine for most private fire equipment collectors. However, this same bulk makes the ladder trucks desirable to the scrap dealers and it is unfortunate, but not uncommon, to see a collector outbid by a scrap dealer on one of those giants.

RESCUE UNITS AND AMBULANCES

Used for rescue operations and medical life saving missions, these units tend to be large vans or conventional panel trucks which have been converted to emergency fire department use. In days gone by, the long-wheelbase Cadillac chassis served as the basis for many ambulances (Fig. 2-5). Today, the short-hooded van has emerged as the vehicle of foremost utility for such purposes. With stand-up room and enough equipment to make a rolling hospital emergency room, they are becoming more common in all parts of the country as firemen in metropolitan, small town and rural departments increasingly accept responsibility for paramedic emergency work.

CHIEF'S CARS

Basically a conventional passenger car with distinctive paint scheme to match the department's trucks, the fire chief's car has been overlooked by many collectors (Fig. 2-6). Most fire buffs prefer the more complex and exotic nature of the specialized fire trucks to the chief's car which is usually little more than a fancy automobile with a small complement of emergency equipment.

However, the conventional nature of the parts and func-

Fig. 2-3. The tanker can sometimes be fitted with a pump like this 1953 Ford/Darley, or it can be equipped strictly for carrying water.

Fig. 2-4. Aerial ladder trucks present unique problems of storage but they also have unique potential for fun at a muster as shown by this 1940 Seagrave performing at the annual Indianapolis, Indiana muster.

tion of the fire chief's car make it probably the easiest of all restoration projects to be undertaken on vintage fire equipment. The earliest fire chiefs' cars were usually powerful luxury models, at least when the community could afford them. Other communities, especially in the post-World War II era, turned the purchasing decision over to the city bean

Fig. 2-5. Ambulance and rescue trucks have been around for a long time, as witnessed by this World War I French Renault army hospital on wheels. You might shake your head at the magnitude of this restoration, but the vehicle was purchased by a museum and will someday be as good as new.

Fig. 2-6. Chief's cars offer the least expensive restoration. Modifications on this 1963 Ford Galaxie 500 would easily make it a proper chief's car with siren, lights and a bright red paint job.

counters who usually bought the lowest priced four-door sedan. Since such economy series sedans are low on the car collector's priority list, a chief's car from the 1940s or 1950s would be a fun, but low-cost project for the fire buff on a budget.

SPECIALTY UNITS

Many fire departments have purchased or created rigs to meet special job needs (Fig. 2-7). These range from a stretched Crosley miniature fire engine for parade use only, to a four-wheel drive Jeep brush fire fighter, to the giant airport disaster vehicles loaded with specialized equipment. Collectibility of these fire engines varies greatly. It is dependent on the individual unit's size, aesthetic appeal and problems the often highly specialized equipment might pose during restoration.

COMBINATION UNITS

It is not at all uncommon for a fire engine to have been built and converted for the purpose of doing several jobs at the scene of a fire. Such trucks are called *combinations*. In the

Fig. 2-7. It is not known what special need the Dolton Fire Department had for this 1941 Diamond T/Darley panel truck pumper, but it is typical of a special rig. Note features such as the front-mounted pump, scroll type hose reels inside the truck with special doors for the hose to emerge and fancy full chrome wheel covers.

jargon of the fire engine hobbyist, a truck that performs two jobs, such as a tanker for hauling water with engines to pump it, would be called simply a *combination*.

A truck that is pumper, tanker and hose carrier would be called a *triple combination*. And a truck carrying ladders or chemicals, plus water, hose and a pump, would be called a *quad*. There are about as many combinations as there are combinations of different trucks, departments and the towns they served.

Each community's special needs were taken into consideration when such units were built. Since combinations are the most common types of fire engines encountered in the collector market and offer a fascinating variety of form and function, they are often chosen by the beginning fire buff for his first restoration project.

However, this same variety of purposes for which the truck was built often means multiple headaches for the collector/restorer. He must familiarize himself with several aspects of the mechanics of fire fighting during the same refurbishing project. Still, when restoration is complete on the combination, or one of its more talented big brothers, it is often everything the fire buff was looking for when he first made the decision to buy, restore and own a vintage fire vehicle.

3

Cost Versus Value

When purchasing a fire engine—whether a project truck in need of ground-up restoration or a ready-made collector's item—the first and most important rule of thumb is to buy what you like. If you like a unit, don't worry about its value or what your friends, neighbors or wife think (up to a point, remember you still have to live with her). The fire engine that you want and will be most happy with is the truck for you. A compromise on this point could turn one of the most exciting hobbies in the world into a time and money consuming disappointment.

INVESTMENTS

Even if you have to pay a little more for the unit than you would like at the time of purchase, the normal process of price appreciation will compensate you in the long run (Fig. 3-1). But, while your investment in a fire engine and its restoration will most likely outpace inflation and such conventional investments as a savings account and the stock market, it is not a get-rich-quick proposition. From purely a hard dollar investment angle, there are many areas of the collector car market that are better bets. If investment or speculation is your only motive, go directly to a classic car auction with a

Fig. 3-1. Remember to be realistic as to what you can tackle. A truck as straight, complete and fully equipped as this American La France should offer no obstacle, even for the novice.

loaded check book; fire engines are not the blue chip stocks of antique motor vehicle collecting.

Another word of caution is probably appropriate here, and while it is basic to all hobbies, it is especially worth repeating in connection with the collection and restoration of fire engines. Do not bankrupt your family, clean out the kids' college fund or cash in your life insurance to finance a fire truck purchase. The potential for financial disaster and your worry about it would take all of the fun out of ownership. An antique fire engine is not the most liquid investment in the world. As a matter of fact, if the time comes when you have to get your money out of the truck in a hurry, it may be impossible. While many of your friends, neighbors and relatives will envy your fire engine, few will be in the position to buy it if and when it comes up for sale. And your chances of selling it to your local fire department are virtually non-existent. Remember, you might have bought it from them in the first place when they obsoleted the truck with a newer model.

There will be more on buying and selling a fire engine in a later chapter, so suffice it to say at this point: don't let your desire to own an antique fire engine obscure your better judgment and common sense. On the other hand, don't let these cautions scare you away. The purchase and restoration of a fire engine need not require that great heaps of cash be laid out at one time. In fact, it is a project that is usually best when it is savored over a long period of time and you are personally able to invest your time and money. And as we indicated earlier, over the long haul that investment will come back to you and the fun you had on the restoration project as well as the joys of owning and using your own fire engine will also be part of your personal gain.

FIRE TRUCK VALUES

The following paragraphs include some basic rules on fire truck values which should prove useful as a guideline.

Custom-Built Engines

The custom-built fire engines (Fig. 3-2) from firms which

Fig. 3-2. Large custom-built apparatus tends to have a higher value than a commercial-chassised vehicle. However, size presents storage problems and an aerial ladder like this 1948 Seagrave owned by Bernallo County, New Mexico, can offer some sales resistance.

specialized in the production of fire fighting apparatus, such as American La France, Ahrens Fox, Crown, etc., tend to be worth more to today's collectors than commercial-chassised rigs like Ford, Chevrolet, International Harvester, etc. This holds true even when the commercial chassis is fitted with a body and other special equipment from one of the custom fire engine builders. To the fire buff buying and restoring an antique fire engine, there is a magic to names like American La France that is just not present in the name of more mundane manufacturers. Therefore, they are usually willing to pay more for that magic.

Combination Equipment

Combination equipment such as pumper/hose trucks and pumper/ladder trucks tend to have and hold more value than a fire truck that performs only a single task. Here the buyer in the collector market is looking for more utility value for each hobby dollar spent, as well as the added visual appeal the combinations usually have over the one-job rigs.

Smaller Engines

The smaller fire engines (Fig. 3-3) while more desirable to the average fire buff from the standpoint of practicality than a hook-and-ladder rig, do not vary greatly in value as compared to the larger units. The difference in saleability between the biggest and the smallest trucks is principally the size of their potential market. There is not a general rule in the antique fire engine marketplace that equates the size of a truck with its dollar value. A hook and ladder might be worth almost as much as a pumper of the same marque, but when it is offered for sale, there will be fewer buyers who have storage space for the 50-foot-long rig.

Many of the desirable, older pumpers, on the other hand, will fit in a conventional garage. The author once owned a 1919 American La France type 75 combination pumper/hose truck/ladder truck. It measured less than 24 feet in length and being an open truck with no windshield, it stood 6′10″ at its highest point—the steering wheel. He was able to keep this fire engine in his apartment garage in Los Angeles. It made for a highly practical collector/truck relationship, though it was unnerving to the elderly lady who lived in the apartment directly above the truck's parking space each time the giant unmuffled T-head engine was fired to life for an afternoon spin around Los Angeles. If that fire engine had been just inches longer or a bit higher, special storage facilities would have been needed, and in a metropolitan area that would have meant high costs which might not have fit into the author's limited hobby budget. When it came time to sell the fire truck, its desirable size became a very important factor to several prospective purchasers and the engine went to a new home very quickly.

Cases have been known where not a single bid was received on an extra-long vehicle like an aerial ladder rig when it was put out to pasture by a fire department. Thus, many giant rigs have been purchased by collectors who think big for surprisingly low bids.

Equipment Stripped Trucks

Fire trucks that are sold with most of the loose equipment removed (as is often the case when buying from a

Fig. 3-3. A small pumper like this American La France combines the best of all the aforementioned attributes: compact size, custom manufacture and drivability. Small chain-drive ALFs like this one rank at the top of collectors' lists.

department) will naturally be worth less than a fully outfitted truck (Fig. 3-4). The original style and type of equipment can be difficult or impossible to replace or duplicate. Sometimes the auxiliary equipment can be purchased separately from the fire department. Occasionally, a department will put up large boxes of obsolete equipment and accessories for bids at the same time an engine is offered.

Exotic Fire Fighting Vehicles

Exotic fire fighting vehicles tend to be more desirable than the average units. Among the most desirable of all pumpers are the giant Ahrens Fox units with the large chrome pressure ball and piston pump mounted on the fronts of the trucks. Oddities like chain drive on American La France trucks and the cowl placement of the radiator on bull dog Macks make a fire engine more popular among hobbyists (Fig. 3-5). Seagrave, American La France and several other fire equipment manufacturers offered V-12 powerplants which were direct descendants of the engines in luxurious Pierce Arrow and Auburn autos. The presence of such an engine in a fire truck adds to its value. Such visual attractions as fancy scrollwork and lettering and large amounts of well-maintained brass and chrome and make a fire engine more spectacular and more saleable, as well.

TRANSPORTATION

One factor to which too few fire buffs give due consideration after they find the truck of their dreams is how they are going to get a 12,000-pound hulk of, perhaps, non-functional fire engine back to their home. Even if the vehicle is fully operational you may have doubts about driving it home.

Obviously, driving the fire engine home is going to be the easiest and most economical solution in most cases. You will want to check the truck out fully to make sure it is, in fact, road-worthy. Check the brakes, water, oil, all the lights and then start the engine and allow it to run long enough to get good and warm so you can check for major leaks or malfunctions. Recently, the author and several friends from the local hobby fire company purchased two antique Seagrave

Fig. 3-4. When most of the equipment has been stripped off a truck, it reduces the value. This 1927 Seagrave pumper is missing hoses, ladders, nozzles, etc., not to mention the original durm-style headlamps that would match the search light still attached.

pumpers from a location about 250 miles from home. The master plan was to drive both trucks home because they were operational.

The first truck was a 1927 model and more fragile than the later-model 1942 pumper. Once all systems were go, both trucks hit the highway. The gas gauges were non-functional on both units but from past experience, it was safe to expect about 4-5 m.p.g. A mileage check on the 1942 later revealed a yield of 4.8 m.p.g. At about the 100-mile mark, the 1927 lost a generator pulley and since we knew we would be driving after dark we knew that the lights would eventually run the batteries down to nothing. Therefore, temporary storage was sought at the home of a fellow fire engine enthusiast. Meanwhile, the 1942 model was driven the full distance, using only a modest amount of oil and a not-so-modest amount of gasoline.

Two weeks later the 1927 was loaded on an open flat bed tractor-trailer truck and transported the rest of the distance. One point not mentioned earlier was that the driving part of the trip was made in December, in open cab trucks through Minnesota and Wisconsin at a time when the average daytime driving weather was 12 degrees. Once again, the enthusiasm of the fire apparatus collector circumvents all logic.

As we chose to haul the 1927 Seagrave the rest of the way, you might also elect this method of transporting your fire engine from its old home to your garage. Such transportation can be in arranged several ways. A check of the suppliers' listings in the appendices will show several trucking firms which specialize in hauling antique motor vehicles. Local truck lines may also be able to help you, especially if you are willing to let the truck be hauled on an open trailer as opposed to a closed truck. Truck rates are generally regulated by the Interstate Commerce Commission's rules which specify that you can expect to pay a substantially lower rate for an open truck than for a closed moving van type of truck. Low-boy trailers, such as are used to move heavy construction equipment are especially good if they can be arranged due to their low-loading height and heavy-duty carrying capacity.

Some towing companies make a habit of towing large

Fig. 3-5. Exotic trucks like this chain-drive Mack Bulldog bring a real premium in the marketplace.

trucks and they can also be contracted to tow your first engine. For shorter trips, this system has been found to be quite acceptable. One towing company once used an item called a *stiff hitch* to tow a 1919 American La France pumper. It was like a very large tow bar. All four wheels of the fire engine rode on the ground and the owners had to ride in the truck to help steer it. It worked well over the short haul but would be impractical for long distance transportation.

For very short trips a very substantial chain, and an even more substantial tow vehicle can be pressed into service. Be sure to check for compliance with local laws and remember that when the truck is not running, items such as power brake boosters, etc., are not operating and it will take both feet to stop the truck.

If you plan to have your fire engine transported on a trailer such as a car hauler, be sure to plan ahead. Most typical car-hauling trailers have a gross vehicle weight (GVW) of 8,500 pounds, including the weight of the trailer which is normally around 1,500 pounds. If you have any idea that your fire truck weighs more than three tons (and unfortunately most of them do) you will probably not want to use such a trailer at the risk of a broken axle or other trailer part, or even an accident. The average pumper usually weighs between 8,500 and 12,000 pounds dry so unless you have an exceptionally small truck such as a Model T Ford, you will be out of luck on most standard car haulers. If you are lucky enough to have access to a trailer with a higher GVW, get out the measuring tape. It will be a disappointing trip if you travel several hundred miles with a car hauler only to discover that the truck wouldn't fit on it anyway. Don't measure only for length, either. Consider width, especially for those trucks equipped with dual rear wheels.

If you get by the weight and size problems the only major obstacle left to contend with is what type of vehicle you will use to tow the trailer. You should use the biggest, strongest, heaviest vehicle you can beg or borrow. I have had good luck with a 454-c.i. V-8 powered Chevrolet Suburban—a station wagon-type truck. Likewise, most adequately powered pick-up trucks would do the job. Most passenger cars are not heavy enough for this kind of work. Even the big station wagons like LTDs and Buicks have suspension systems that are entirely too soft.

It is possible to trailer a fire truck, but just make sure that you consider all factors before attempting it.

Above all, don't be too disheartened by shipping and hauling problems. I recently read of a man from Alaska who bought a large American La France fire truck in Wisconsin and drove it to Washington state. From there he had the truck loaded onto a boat and shipped to Alaska. It just goes to show you that where there is a will, there is a way!

4

Major Marques: A Brief History

When you actually begin the process of searching for the fire engine of your dreams, as outlined in the first chapter, you will most likely do so with a stereotyped image of the ideal fire engine firmly fixed in your mind. If you are in your twenties or thirties, that perfect classic fire engine could well be the round-front, mid-engined V-12 American La France that was the staple of many fire companies in the post World War II era. If you are a little older and have lived in a major city, you most likely have vivid memories of the big Ahrens Fox units with their giant chrome pressure ball out front. Maybe a Detroit man thinks immediately of a Seagrave sedan pumper, while the rural collector's image of an antique fire engine revolves around a humble Model T Ford with a simple gear driven pump.

Many potential fire engine collectors, however, just happen onto an available truck that strikes their fancy. In any case, before making such a major decision, the fire buff would be wise to gain a working knowledge of the different fire engine manufacturers, the individual characteristics of their equipment and any potential problems peculiar to the various marques. In the next chapter you won't learn all there is to know about each of the major makes of fire fighting equipment, but you'll pick up enough so that you won't be completely in the dark when you enter a conversation with other collectors. Be careful, though, because in the collecting and restorating of vintage fire engines a little knowledge is, indeed, a dangerous thing. Use the information in this book to make a decision on what make and model fire engine you want for your own and once you have zeroed in on those specifics seek out books and other sources of information that will make you a real expert in your chosen field. There are many fine books and many manufacturers are still in business: You will be surprised at how cooperative they will be in helping you with your restoration.

AHRENS FOX

Ahrens Fox Fire Engine Company of Cincinnati, Ohio, was a direct descendant of the Ahrens Manufacturing Company. Also of Cincinnati, it was a firm that was well known for its steam pumpers as far back as the turn of the century.

Later, in 1910, Ahrens Fox introduced a battery-powered *Continental* fire chief's car which also carried fire fighting chemical equipment. By that time fire departments were using fewer and fewer steamers. By 1911, Ahrens Fox, as the company had become known, had developed its first gasoline engine vehicle. Even the very first Ahrens Fox pumper featured the front-mounted piston type pump that was to become a trademark for the firm and which now makes the units highly prized and dependable collectors' items (Fig. 4-1). The earliest units featured two domed pressure balls on top of the pump. By 1913 the company had become an industry leader and introduced the *Booster system* that all but obsoleted the traditional soda and acid chemical tanks that had been standard equipment for years. Just a year later, the fast-paced firm developed a line of its own T-head type power plants to replace the Herschell-Spillman engines it had been purchasing and installing in its trucks. It was also in 1914 that Ahrens Fox introduced its new *K model* featuring the single round dome pressure ball over its forward-placed pump. It was this design that made Ahrens Fox fire engines so effective and so visually unique for many, many years.

Ahrens Fox entered the aerial ladder truck business

Fig. 4-1. Without a doubt the most sought-after antique fire engine of all is the famous Ahrens Fox pumper with its front mount piston pump and high pressure ball.

commercially in 1923. These rigs used a unique air-lift system that had been tested on a prototype unit many years earlier. It was also about this time the single spherical ball pressure chambers began to be manufactured in two pieces. A story says that the old craftsman who had been hammering the balls out of a single piece of metal was getting on in years so the company hired an assistant who was to learn the art of metal working from the old man. However, the veteran craftsman apparently viewed the young assistant as a threat to his job and refused to teach him anything. A short time later the old man died, taking the secret of his skill with him to the grave. From then on, Ahrens Fox pumpers carried a two-piece pressure ball, held together with a horizontal strap. Whether or not this story has any validity is still debated among fire engine collectors today, but it is a fact that pumpers equipped with the single-piece ball brings a premium over those with the two-piece unit.

By the mid-to-late 1920s, Ahrens Fox was manufacturing *Rotarystyle mid-ship pumpers* as well (Fig. 4-2). While these are good units, they have yet to match the mystique that the firm's piston pumper units enjoy with the collectors.

By the early 1930s, Ahrens Fox began to offer a line of smaller trucks. These smaller trucks were aimed at fire departments in small towns. They were offered in the full range of the firm's styles, with pumpers limited to outputs of about 500 gallons per minute (g.p.m.). The bigger pumpers began to sport a more streamlined styling and by the very late '30s, a few were even featuring headlights molded into the fenders. The styling on both basic types of pumpers—the mid-ship as well as the piston driven—began to get quite massive and very impressive as the Ahrens Fox company moved into the early 1940s. By the late '40s, the 'Foxes had taken on a more modern, but more homely, boxy styling on the mid-ship pumpers and ladder units.

The end of the positive output piston pump for Ahrens Fox came in 1952. The centrifugal pump mounted mid-ship was less costly to produce and had made great gains in efficiency over the four decades the firm had produced the front-mounted piston units.

In 1953, the C.D. Beck Company of Sidney, Ohio, pur-

Fig. 4-2. This Ahrens Fox from the John Kress collection is typical of Foxes with midship pumps and ladder trucks.

chased Ahrens Fox. The manufacturing operations were moved from Cincinnati to Sidney. Styling became even more modern, and even more box-like in appearance. Their future as collector items was on the wane. The company introduced a new line of cab-forward trucks in 1955, but by 1956 Ahrens Fox had produced its last pumper.

In 1961, Richard C. Nepper of Cincinnati bought the remaining parts, records and the good will of the name Ahrens Fox from Beck. He is still in business today, helping to preserve the famous name and the Ahrens Fox legacy as a collector item.

Besides the guidelines laid down in an earlier chapter which concerned purchasing a vintage fire engine, the collector who has decided on an Ahrens Fox should consider a couple of recommendations which are specifically applicable to the marque.

The most popular Ahrens Fox fire trucks, from the standpoints of demand in today's market and probable future resaleability, are those with the early one-piece ball pressure dome units (Fig. 4-3). However, even with the two-piece pressure ball, any Ahrens Fox with the front-mounted piston pump is desirable as a collector item to any fire buff. That popularity, naturally, is reflected in the price structure for these units on the market. It is not uncommon at this time to see unrestored Ahrens Fox fire engines sell in the $2,500 to $5,000 range, with $4,000 a very common asking price. A very solid original 1924 piston pumper sold recently at a Tulsa, Oklahoma, collector vehicle auction for a respectable $4,800. In fully restored condition, $15,000 to $20,000 is the frequently encountered asking price. We recently saw a late '30s Ahrens Fox advertised with an asking price of $38,000, although there seemed to be few interested parties at that price.

Ahrens Fox fire engines are large by nature, but their popularity among collectors generally defies what was said earlier about the greater marketability of compact rigs. The name and mystique which surrounds the Ahrens Fox fire equipment seem to counteract rules which normally tie size and value together.

AMERICAN LA FRANCE

What is probably the greatest and best known name in fire fighting equipment—American La France—traces its ancestry to the American Fire Engine Company which manufactured horsedrawn steam pumpers and hose carts.

The International Fire Engine Company became American La France's closest antecedent with the manufacture of self-propelled steam-powered hose and chemical cars. The year 1903 marked the formation of American La France as a conglomerate of nine different manufacturers, including the two aforementioned. The firm moved to Elmira, New York, where it remained for many years.

The company had been making steamers since 1897 and ladder units since 1903, with a *spring-loaded ladder* unit being an industry pacesetter. The company did not enter the gasoline powered era until 1910, although American La France did build a prototype in 1907 and a few units on Simplex passenger car chassis before 1910. Simplex passenger cars are among the most desirable automobiles from this era among antique auto collectors, so it is safe to assume that if a fire buff

Fig. 4-3. A close-up of a dirty, but still valuable, Fox with the now-famous one-piece chrome pressure ball.

finds one of these fire vehicles, it would be a real prize. These early trucks featured the chain drive which was to become ALF's trademark for many years.

If anything could be said for American La France, it was that they offered diversity. In the early 'teens they offered a unique gas/electric aerial ladder truck on which gasoline engines powered General Electric locomotive engines that drove the front wheels on the big rig. The firm also offered gasoline powered front-drive two-wheeled tractor units to pull ladder trucks and convert old horse-drawn steamers. The company's line also included conventional four-wheel rear drive aerial ladder trucks with a rear tiller, as well as pumpers, hose cars and chemical trucks. It was not uncommon for American La France equipment to be found on commercial chassis built by other manufacturers, such as Ford's Model T, Chevrolet, etc. ALF has continued this tradition to this day, offering everything from the most exotic custom-built equipment to simple equipment fitting on standard truck chassis for fire departments with less needs and smaller budgets.

The basic designs and classic square-edged hood and radiator configuration of the 'teens carried over on ALF equipment of the '20s (Fig. 4-4). The aforementioned diversity also carried over. American La France became one of the largest, if not the *the* largest, manufacturers of fire-fighting apparatus in the industry. The general look and design of its trucks from the 'teens and '20s have made them the most-sought after of all units by American La France collectors.

The 1930 American La France Metropolitan was the fire truck of nearly every fire chief's dreams. Large, but compact, the Metropolitan was a triple combination pumper of the first order. It could pump 1,000 g.p.m., carried a 40-gallon chemical tank, 1,200 feet of 2½-inch hose and a modest selection of ground ladders. This truck was to set the pace for combination trucks for many years to come and today represents the ideal truck for the fire buff looking for a multi-purpose fire engine with the look and feel of a big rig, but without the mass.

In 1935, American La France introduced a long-nosed, low slung series called the 400. Most of these trucks were powered by V-12 engines. Adding to the length of the cowl, the pumps were mounted just behind the fire wall. Only built for three years in any quantity, many collectors consider the 1935 ALFs the most attractive ever turned out by that manu-

Fig. 4-4. American La France has been a leader in ladder trucks since 1903. This late 'teens or early 1920s' straight-chassised ALF is now hobby fun for a collector.

facturer. In direct contrast, the 500 series introduced in the late '30s and the 600 series put into production shortly thereafter were among the least appealing trucks built. Large and ponderous, the bulky 500 and 600 series fire trucks earned all sorts of nicknames from the men who manned the rigs. One Wisconsin fireman described his department's 600 series pumper as driving "like a big red submarine" through crowded city streets. He also remarked that the truck was especially clumsy to drive and even seeing out of it was difficult.

American La France introduced its 700 series cab-forward design in 1946. It revolutionized fire engines and continues to affect their design to this day. The truck was incredibly compact, yet carried a full complement of equipment. Styling was so perfect that ALF (and many other manufacturers) still utilize the basics of the design today. To the post-warfare buff, this is the truck that embodies the classic fire engine ideal of the era. Children's toys, coloring books, even your own stereotyped image of what a fire engine looks like are, in fact, traceable to the American La France series 700 line. Some collectors, though, shun the truck today because it is still too contemporary. This idea has kept the market value of the series fairly low. It is great, however, for the collector who has decided that an ALF series 700 is the fire engine for him! Parts are still plentiful for these V-12 powered brutes which helps to make this truck one of the best buys in the field of vintage fire-fighting vehicles.

Even American La France's most recent giants still carry the 700 series styling, although they are more rounded and refined. ALF also continues to pioneer in the areas of fire apparatus engineering and technology. In recent years they have been developing turbine powered fire engines, turbo-charged diesel powered trucks and every other combination of power train available to modern trucks. American La France is still the scale by which custom-built trucks are judged (Fig. 4-5).

There also is some specific advice which should be of interest to the collector looking for an American La France fire engine. The early 'teens and '20s ALF trucks with chain

Fig. 4-5. Seen here at the Indianapolis fall muster, this late 1940 American LaFrance open cab pumper is the classic fire engine design pioneered by ALF.

drive are the most in demand. Current prices range from $2,500 for a restorable truck to $10,000 for a unit on which the restoration has been completed. The Metropolitans and 400 series units from the '30s are magnificent, but larger and less antique looking, so they command about 10 percent less. Since 700 series trucks have not yet caught on to the degree which we expect them to in the future, they are the best bargain of the lot. They start under $1,000 and seldom go over $4,500. The homely series 500 and 600 trucks hit bottom as well, under $1,000 in poor condition to just about $4,000 at the top. It is our opinion that the 700 American La France fire engines are the best investment if the collector is looking for dollar growth. With its modern classic design and traditional engineering superiority, it is sure to gain popularity rapidly as a new generation of fire buffs enters the collectors' market.

Also a boon to the American La France hobbyist are the several support groups which include a very active program of cooperation from the factory. They actively promote the restoration and preservation of their trucks. For further information, see the directory of organizations and restoration aids listed in the appendices.

BUFFALO

The Buffalo Fire Extinguishing Manufacturing Company had been in the chemical extinguisher business for many years prior to 1920 when it was decided to expand the operation into the area of building complete fire trucks on commercial chassis (Fig. 4-6). By 1930 Buffalo had gone into the production of its own custom chassis trucks and became well known for large, rugged equipment. New styling introduced by Buffalo in 1934 gave the firm's trucks the appearance of being much more streamlined than the bulk of the competition. Carrying the streamlining one step further, Buffalo introduced a new line of fully enclosed pumpers in 1939. A unique feature of that series is the pump placement. It is behind a set of panel doors and out of sight when not in use. It would be interesting to talk with firemen of that era who may be familiar with the closed Buffalo pumpers to see if the fully enclosed pumps presented any obstacles to getting hooked up

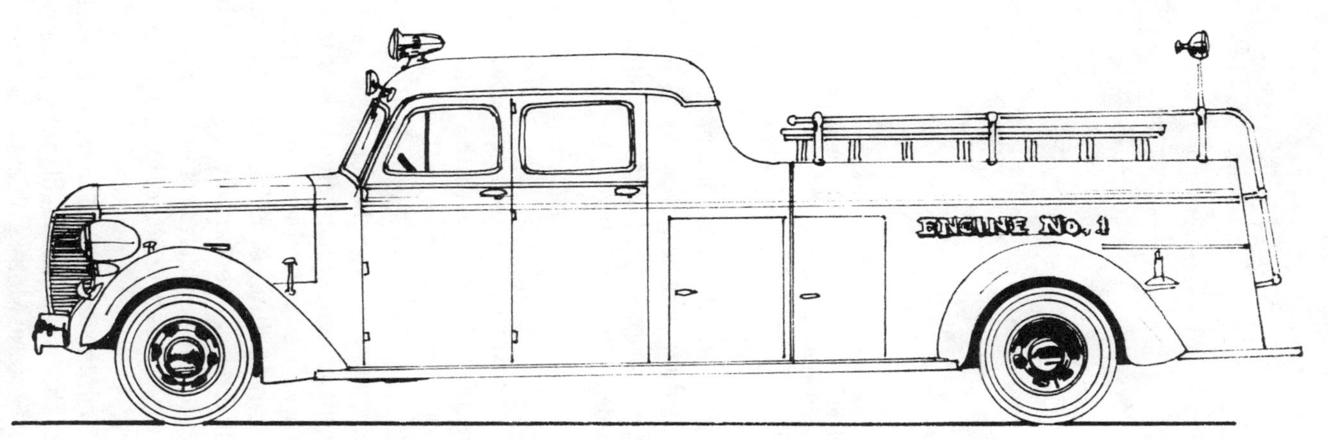

Fig. 4-6. Buffalo was well known for the large, long closed cab models built in the post-war era, like this 1941 750 g.p.m. sedan cab pumper.

Fig. 4-7. Chevrolets have been used for commercial-chassised fire engines since the very earliest days of motor vehicles. An example is this late 1920 open cab, front-mount pumper.

quickly for fire work. Their comments on the extraordinary length of most Buffalo fire engines would also be interesting. After a long, illustrious stand in the fire apparatus business, Buffalo closed its doors in 1948.

CHEVROLET

Chevrolet never built a fire engine per se; however, we include them here because with the possible exception of Ford, Chevrolet produced more chassis for commercial body builders than any other manufacturer (Fig. 4-7).

Chevrolet autos got into the act as early as the late 'teens as chief's cars and light-weight chemical trucks. Many early Chevrolet trucks were fitted with fire equipment by Barton, Hale, American La France and Boyer. By the late 1920s, Chevrolets had become especially popular with commercial body builders for the production of fire engines. As time went by, the Chevrolet six-cylinder engine became a hallmark of dependability.

Up through the thirties, Chevrolet-powered apparatus remained on the small and light side, mostly due to the preference of General Motors to leave the heavier truck lines

Fig. 4-8. One of the most sought-after fire engines in this Model T Ford. The Ford name, small size and ease of getting parts contribute to the popularity. This small pumper was built by Waterous in November of 1919.

to the GMC nameplate. As the forties dawned, however, Chevrolet began building bigger and heavier trucks, including some attractive cab-over-engine designs. With this advent came more and heavier commercial chassis Chevrolet apparatus. By the early sixties, Chevrolet and GMC had almost identical lines of trucks. Manufacturers like John Bean used both brands of chassis, though in some cases still leaned toward GMC for heavier components. By the mid sixties Chevrolet had an excellent tilt cab truck that lent itself well to commercial chassis application. It is ironic that though Chevrolet shared so many components with GMC, more manufacturers chose to link their name to GMC units due to a reputation for heavier trucks.

Today we see many manufacturers using Chevrolet chassis as the basis of mini-pumpers, brush trucks, ambulances and, of course, larger full-size pumpers.

FORD

Without a doubt, Ford has provided more commercial chassis to fire equipment body and apparatus builders than any other single manufacturer (Fig. 4-8).

Starting as far back as the Model T which was introduced in 1909, Ford cars provided the basis for many chemical cars and chief's cars. Ford's truck chassis of that era was referred to as the TT, and both the *T* and *TT* chassis were stretched in every direction to build apparatus. American La France and Waterous were some of the earliest manufacturers to use Ford chassis for their equipment. Boyer and Pirsch added their names to the ranks of Ford users and the current manufacturer, Pierce, had its beginning building commercial truck bodies on Model T chassis. Additionally, the Model T was so lightweight and versatile that many fire departments used the engine and radiator unit to power portable pumpers and light generating equipment.

In 1928 Ford introduced the Model A passenger car and the companion Model AA truck. Once again, these Ford vehicles were an immediate hit with manufacturers. The Model A (Fig. 4-9) combined the efficiency of the Model T and its renowned dependability with a much higher operating speed for the Model A. The new commercial chassis became widely used throughout the industry virtually overnight. Today the Model A Ford is the most widely collected antique automobile, with hundreds of thousands still on the highways. It is no wonder Model A Ford-chassised fire apparatus has a double appeal to fire engine collectors and to Model A fans in general. It is this type of popularity that makes Model A-based fire equipment super collectible today as well as fairly high priced.

In 1932 Henry Ford brought out his soon-to-be-famous flathead V-8 engine. By the mid thirties it had gained acceptance with almost every commercial builder in the business. The flathead became a legend in its own time for power and reliability. So much so, in fact, that Henry fell behind in the engineering race toward the modern overhead valve engine. It was not until Henry Ford's death in 1947 that Ford engineers really got the go-ahead to produce a modern powerplant (Fig. 4-10). Lincoln was the first Ford product to get the new type of engine in 1952 and commenced to win a couple of Pan

Fig. 4-9. The only small truck in greater demand than the Model T Ford is the Model A, like this 1929 pumper. Dependability and drivability make the Model A's an even better bet for the collector than the earlier Fords.

American road races in Mexico in following years. Ford introduced the OHV engine in its cars and trucks in 1954. Early models produced the same displacement as the flatheads they replaced, but offered more efficiency. Early OHV Ford engines had a reputation as oil burners, just as the flathead did in 1932, but this was soon cured in both cases to develop into a line of strong, powerful and dependable powerplants.

Today the Ford Budd-bodied cab and chassis unit is the basis for almost every commercial truck being built. This highly versatile chassis and cab was introduced in 1957 and continues to this day. Even Mack used the Budd body for a couple of years. The C series Ford cab and chassis could be the most widely used commercial chassis ever built. Today, several manufacturers are also making use of the large L series Ford and Louisville lines as the link between the medium-size C series commercial-chassis rig and the huge custom-chassis pumpers. The L series is just the right truck for the job and offers communities on a budget big truck performance at a commercial chassis price.

FOUR WHEEL DRIVE

Four Wheel Drive (FWD) was founded in Clintonville, Wisconsin, after its founder, Otto Zachow, broke away from Oshkosh Truck down the road. Known for durable 4x4 trucks, the FWD made an ideal truck for rural use and was an almost immediate hit. Chosen by the United States Army for a huge contract of military vehicles in the first World War, FWD made quite a name for itself early in its career. The rugged Model B was the truck the company was built on. Thus, it is easily understandable why the Model B was the basis for many early 4x4 fire engines. FWD built its own apparatus, as well as offering its chassis to other builders.

The FWD had a reputation built in the war and in civilian service of being a mountain goat capable of handling any terrain. Having driven a Model B of 1919 vintage, we can testify that these vehicles seem almost unstoppable. The Model B has a top speed of 14 m.p.h. and seems willing to do that on flat pavement or straight up a brick wall.

As the FWD reputation spread, city fire departments got

Fig. 4-10. A late example of a very common and popular commercial chassis, the Ford F-7 of 1950, is shown with W.S. Darley Co. equipment.

the word and began to realize how beneficial the four-wheel drive feature could be on city streets under wet or snowy conditions. The rugged construction for which FWD trucks were fast becoming known was also a major selling point. By the mid thirties FWD was providing pumpers and aerial ladder tractors to many municipalities. Paralleling this line of equipment building, FWD had also become one of the biggest manufacturers of snow removal equipment.

In the late forties and early fifties FWD delivered a large quantity of its uniquely-styled open cab pumpers throughout the United States and Canada. These trucks were very large and the engine was located slightly ahead of the front axle. This design was continued on FWD units well into the sixties. In 1965, FWD purchased Seagrave and moved its operations to the Clintonville facility. Both FWD Tractioneer trucks as just described were continued with the cab forward design of Seagrave. At the same time, FWD was providing Pierce Manufacturing in nearby Appleton, Wisconsin, with many components and chassis. FWD and Seagrave are still thriving in Clintonville, a major force in the fire apparatus industry (Fig. 4-11).

GENERAL

It is not known exactly what year the General Manufacturing Co., of St. Louis, Missouri, began building fire apparatus on commercial chassis. It is known that it was well into the twenties (Fig. 4-12), as evidenced by the 1928 Pierce Arrow-chassised trucks. General did, however, begin manufacturing its rugged line of custom chassis trucks in St. Louis in 1932, and soon developed a reputation for its big, heavy equipment, similar to Buffalo. Also like Buffalo, General adopted a new streamlined design in 1934, including a split V-windshield.

General soon became well known for unusual and strikingly beautiful fire trucks. For example, in 1937, General manufactured a closed cab pumper which carried a complete Ford Deluxe Club Coupe body for the back of the truck. The Ford auto body was beautifully integrated into the lines of the truck. Also, by this time, General was manufacturing trucks in

Fig. 4-11. FWDs have not caught on well among collectors, but they were extremely popular with fire departments where the four-wheel drive system came in handy for snowy or muddy roads. This early 1930 FWD awaits restoration in central Wisconsin.

Detroit. By 1938 the firm had developed a striking line of streamlined custom trucks and an even more unusual line of commercial chassised trucks, all based on V-12 Packard automobiles. The Packard/Generals have to be the "classic" fire engines of all time, featuring the styling of the Packard fenders, grill and hood with the appeal of a fire engine. Even details such as large V-12 emblems on each hubcap add to these attractive brutes (Fig. 4-13). By the late thirties, General Fire Truck Corporation of Detroit had become well known for very large, streamlined closed cab and sedan cab pumpers, plus a full line of other fire fighting apparatus. Late thirties grill styling of the Generals is reminiscent of Cadillacs of the same era.

INTERNATIONAL HARVESTER

International Harvester (Fig. 4-14) of Chicago is one of the oldest vehicle manufacturers in the United States. Its famous line of high wheeled buggy-type auto wagons goes back well into the pre-teen era when I-H manufactured a few of its high wheelers in Akron, Ohio, around 1909-1911. Today I-H has production plants all over the country, with major centers of activity in Fort Wayne, Indiana, and Springfield, Massachusetts.

There is little doubt that even the very earliest International Harvesters were the basis for at least a few fire trucks, especially in rural areas where the high wheeler chassis would have provided maneuverability on muddy back roads.

By the twenties, I-H was well known for its line of trucks using the shovel-nose design, almost identical to the Kelly-Springfield trucks. Quite a few fire trucks were built on these chassis. On such trucks, the radiator sat behind the engine, similar to the Bulldog Macks of the same vintage. Today a shovel-nose I-H would be quite a prize for a collector, especially if it is a fire engine.

Through the thirties, forties and fifties, I-H remained a major manufacturer of chassis for the commercial body builders. Even before the firm discontinued light duty truck construction in the mid seventies, the two-ton I-H 4x4s were made into many good brush fighters.

Fig. 4-12. Many GMC chassis were used by the various fire apparatus manufacturers. This finely restored twenties-era GMC features ladders chemical tanks and front-mount pump. This well-equipped little rig saw much action in a rural Wisconsin volunteer fire department.

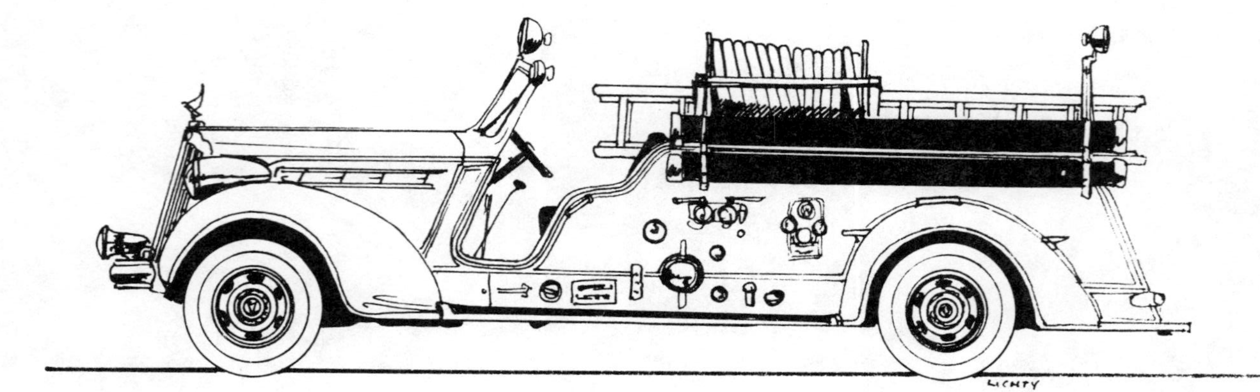

Fig. 4-13. General was known for its fine custom fire apparatus, but few can compare to the General/Packard V-12 hybrids such as this classic styled 1938 750 g.p.m. pumper.

Fig. 4-14. International Harvester was also one of the pioneers in offering commercial chassis to fire apparatus builders. This 1916 chemical-hose car would be a prime example and highly desirable today.

International Harvester is also directly connected with Hendrickson, a component truck manufacturer in the Chicago area. Hendricksons are famous for heavy duty over-the-road trucks. It is little known, though, that these Hendrickson chassis are the basis for many custom-built fire engines as well.

As listed in the club directory in the appendix, the International Truck Restorers club is a good group for activities beyond fire musters. International trucks seem to have a good supply of parts available to the restorer and they make handsome rigs when restoration is complete.

MACK

Mack Trucks of Allentown, Pennsylvania, began an illustrious career in the fire engine manufacturing business in 1911 with the supplying of a Mack motor pumping engine to Bala Cynwyd, Pennsylvania. By the mid-teens, Mack had already established itself as a leader in the production of fire apparatus and trucks in general (Figs. 4-15 and 4-16). The legendary Bulldog Mack design was soon well known all over the world from military use in World War I and on the home front as dependable fire engines. The Bulldog Macks carried the radiator at the rear of the engine, creating a shovel-like hood that was to become a Mack trademark through the 1930s. The sound of the big hard rubber-tired, chain-driven Mack fire engine, or even a cola truck, was exciting for the youth of the day. Today these same Bulldog Macks are some of the most desirable and sought-after apparatus by collectors. So much so, in fact, that the majority of remaining survivors reside in various museums around the country.

Mack also built many radiator-forward trucks with shaft drive. They proved equally dependable, but there was something about the combination of the growl of the big T-head engine, the hard rubber tires and the rattling of the chain drive that gave a certain romance to the big Bulldog.

Since Mack manufactured its own fire-fighting apparatus, it seldom supplied chassis to commercial manufacturers. If a fire department wanted a Mack fire engine, they bought Mack all the way.

The conventional Mack trucks of the forties became something of a classic of fire engine design. Thousands of these trucks were sold and there is hardly a boy of the era who does not remember an early fifties Mack fire engine. Tootsie and Smith-Miller added to the mystique of this classic design by building toy replicas of all sizes.

By the mid fifties the Model B Mack had replaced the earlier model with similar, but modernized styling, such as headlamps enclosed in the fenders. The famous Mack Thermodyne 6-cylinder engine powered most Model B Macks, although some came equipped with the giant Hall-Scott powerplants. Also in the late fifties, Mack borrowed the classic Budd body used on the Ford C series cab forward trucks for a few of its own vehicles, but soon abandoned this tilt cab for one of its own manufacture.

Mack was responsible for building what could be the largest displacement fire pumper on earth for the city of New York in 1965. The giant tractor trailer rig is towed by a Mack

Fig. 4-15. Considered a classic design by many, Macks like this 1952 1,000 g.p.m. pumper were the pattern for many toy fire engines, including the famous Smith-Millers.

Fig. 4-16. Successor to the Mack shown in Fig. 4-16, but no less attractive is this 1965 1,000 g.p.m. pumper from Albuquerque, New Mexico. It is a likely candidate for a collectors' item when it is finally retired.

cab-over-engine, over-the-road tractor. The trailer carries an 18-cylinder, 2400 h.p. Napier-Deltic engine driving a DeLaval six-stage pump at the rate of 8,800 gallons per minute, drawing from as many as eight fire hydrants at one time. To accompany this monster in New York, Mack also built a tractor trailer hose truck and water cannon that can shoot 10,000 gallons per minute while being fed by a quartet 2,000 feet of 4½-inch hose lines. Not a bad rig for a small town!

To this day Mack remains a leader in the fire apparatus industry and the trucks the firm currently produces will be as collectible some day as the early units are today.

MAXIM

Maxim Motor Company (Fig. 4-17) of Middleboro, Massachusetts, was started in 1914 when its founder built a Thomas Flyer-based automobile hose car for his home town. Working from the basis of sound design and craftsmanship paid off as the firm still exists in the town of its origin. By 1916

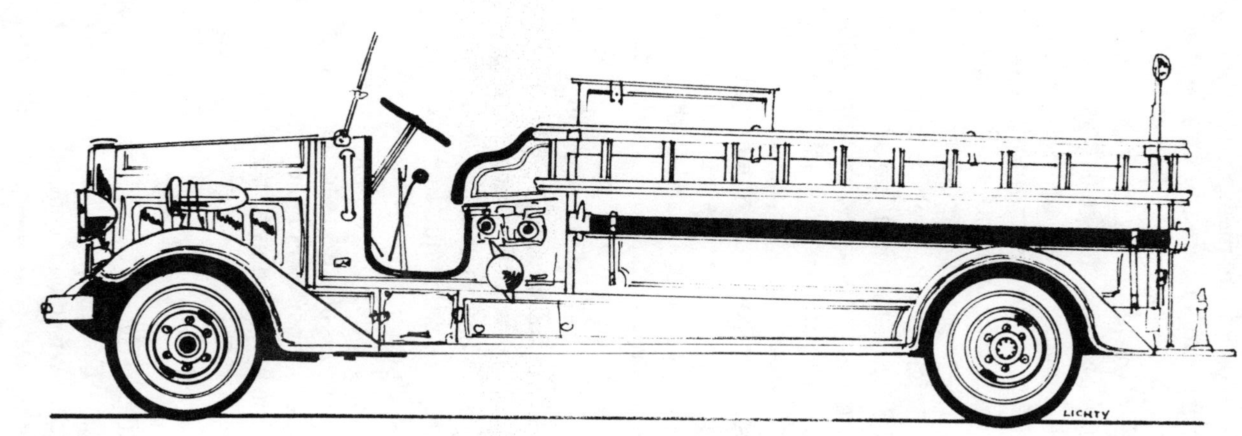

Fig. 4-17. Maxim was another early manufacturer of massive custom equipment as indicated by this 750 g.p.m. pumper.

Maxim had built its first custom-chassised truck. These trucks eventually evolved into the famous M series of 1921, a rugged piece of apparatus offered in a variety of sizes. For 1932, Maxim restyled its complete line of trucks by placing doors in the sides of the hood—an innovation that would be an identifying mark on Maxims for years to come.

Beyond that, Maxim carried basically very conservative styling into the late 1930s, when most of the rest of the industry was swinging towards streamlining. Even into the forties, as Maxim developed more modern cabs, the radiator treatment was straight twenties-era in appearance.

Yet by 1952 when the firm introduced a short aerial ladder truck using a Magirus ladder unit, the grill design had changed to a very attractive rounded unit that would become the marque's styling trademark for the decade. The new grill, matched with nicely rounded fenders and a V-shaped windshield, gave the new Maxims a grand look. As the firm moved into the sixties and expanded their dealings with budget conscious fire commissions, they offered Maxim-bodied commercial-chassised trucks. Though still heavy duty Maxim cab units, they were mounted on big Ford drivetrains and those of other manufacturers.

Today Maxim trucks have reverted to a more conservative styling. It is a bit boxy and definitely conventional looking; however, they are still being turned out in record numbers at the Middleboro plant.

As to the collectibility of Maxims, I prefer the chrome-laden rounded-nose jobs of the late forties and fifties, but that is strictly a personal choice. For the collector looking for a modern driving truck with older looks, the late thirties Maxim might be just what the doctor ordered.

OSHKOSH

Oshkosh, like FWD, has based its reputation on durable, rugged 4×4 type trucks (Fig. 4-18). While FWD in neighboring Clintonville, Wisconsin, concentrated on Army trucks and fire apparatus, Oshkosh was primarily developing off-the-road vehicles, snow plows and other tough specialized equipment, with some military units. We have seen photos of

Fig. 4-18. Oshkosh's forte is giant off-road trucks, 4×4s, construction equipment, snow plows and military vehicles. It is no wonder, then, that when they team up to build a fire truck it is a monster like this 1959 Oshkosh/Pirsch 1,250 g.p.m. 4×4 pumper. This truck would be a very unlikely candidate as a collector's item due to its large size.

Oshkosh-based equipment dating back to World War I service days on which the Army 4×4 style chassis is almost identical to FWDs of the same era.

Typical Oshkosh trucks of today include modern-day behemoths for airport crash services and big off-road rigs for use in the Arabian desert.

Oshkosh never produced the volume of fire trucks of manufacturers such as American La France or Ford, so consequently there are not many around today and it is a rare treat to see a restored Oshkosh.

PIERCE

Pierce Manufacturing Company Inc. is another well-known Wisconsin fire apparatus builder. The firm began in Appleton, Wisconsin, in 1913 by building commercial truck bodies. Then, by 1917 when the firm was incorporated, it was producing commercial bodies of many types for Ford Model T chassis.

Pierce built its first piece of fire apparatus in 1939 and has built on that to become known around the world as one of the industry leaders in the manufacture of custom-chassis fire trucks. So much so, that in 1974 it landed a contract with Saudi Arabia for what could well be the largest single order for fire apparatus in history—740 units including emergency trucks, pumpers, mini-pumpers and heavy duty rescue trucks. It took the firm three years to fill the orders in addition to their more conventional orders.

Pierce entered the custom apparatus field in 1979 with the Pierce-Arrow, a full size custom pumper utilizing the same quality the classic name implies (Fig. 4-19).

It is not uncommon to see Pierce-equipped trucks in collector hands based on Ford, Dodge and GM chassis.

PIRSCH

Peter Pirsch & Sons of Kenosha, Wisconsin, is one of the oldest fire apparatus manufacturers still in existence. They built their first White-chassised truck in 1916, although the firm was established in 1857.

Pirsch built its first custom pumper in 1926, with a

Waukesha powerplant as the base. In 1929, Pirsch introduced several new models including its first 1,000 g.p.m. pumper (Fig. 4-20). This was followed three years later with an even more extensive line of 15 series trucks.

By the later part of the thirties, Pirsch already had adopted a strictly modern look with a concealed radiator shell and grill on some models. By the mid forties, the familiar Pirsch hood line and grill had been developed. This hood line and grill eventually became the trademark of the firm.

Pirsch made wise use of a complete range of cabs from custom-built sedan cabs to commercial cabs outfitted with Pirsch grills. In most cases it was hard to determine the original manufacturer of such commercial rigs as the camouflage job had been so well executed.

Today Pirsch equipment is in high demand among collectors, especially the custom-built units. Since the Kenosha-

Fig. 4-19. Pierce now offers its own line of Pierce-Arrow custom-chassised trucks. It also continues to build many commercial/component type trucks, as well, like this 1974 Pierce/Hendrickson/Morita 135-foot aerial ladder truck for Chicago.

Fig. 4-20. This beautifully restored 1928 Pirsch 500 g.p.m. pumper from LaSalle, Colorado, is a prime example of what to look for in a fire engine: well equipped, modest size and plenty of eye appeal.

based firm is still in business and producing fire equipment to this day, many fire fighters have a strong allegiance to the marque.

SEAGRAVE

Having built its first truck in 1908, Seagrave is one of the oldest manufacturers in the fire apparatus business. A custom builder from day one, Seagrave soon developed a full line of four- and six-cylinder engines after testing several component powerplants available at the time. By 1909 Seagrave had expanded from building just pumpers to its first tractor-drawn aerial ladder truck. Around this time the firm was also experimenting with air-cooled engines. It was also in the preteen era that the city of Pasadena, California, bought a CA-40 Seagrave chemical car which has become famous as a regular feature in the Rose Bowl parade each year. By 1911, Seagrave had again expanded its line, adding several models which had the engine moved ahead of the driver's compartment rather than under it. One brute of a pumper in the new line-up had a 1,000 g.p.m. pumping capacity, far ahead of its competition at the time.

Seagrave continued to be an innovator in the industry and in 1915 introduced the first auxiliary cooling system to maintain constant temperature control when pumping. Seagrave joined the modern world in 1922 by adding shaft-drive models to its selection and offering both wood and steel wheels. By the next year, the Columbus, Ohio, manufacturer had a full range of pumpers ranging from the small 350 g.p.m. *Suburbanite* to the giant 1,300 g.p.m. *Metropolite.* In 1929 Seagrave offered four-wheel brakes as standard equipment on its basic line of pumpers. Seagrave styling by this time was quite attractive, with large chrome-plated radiator shells and large drum-style headlamps.

In 1935 Seagrave made many improvements by restyling the majority of its trucks with a V-shaped grill and more modern fenderlines that would remain similar on Seagraves for many years to come. It was also in that year the firm introduced a Pierce-Arrow-based V-12 engine in its trucks that became a standard for nearly 30 years. Since parts for this engine are still available through the Seagrave factory (now a part of the FWD operation in Clintonville, Wisconsin), many Pierce-Arrow owners solicit the factory for internal engine parts to keep their antique autos running in addition to the Seagrave collectors who need the same parts to keep their fire engines operating.

By World War II, Seagrave had reached a styling pinnacle of a rugged, classic look, with a sloping rounded radiator shell and V-shaped windshield. Open cab models carried half-doors with angled windows that kept out the chilling wind at almost any speed, yet provided top visibility. By the mid 1950s, Seagrave had added a more squared-off frontal treatment, with the siren mounted directly in the middle of the hood. Also adding to the styling of the newer models was the addition of several bold horizontal bars between the headlamps. In 1958 they added dual headlamps to basically the same design, and by the early sixties, the firm was a leader in aerial ladder trucks utilizing several innovative types of apparatus.

In 1963, FWD of Clintonville purchased Seagrave and slowly began to wind down affairs at the Columbus, Ohio, facility. It is interesting to note that during the transition they built trucks in both towns before finally putting the Seagrave into full production right alongside the FWD line in Clintonville in 1965. And to this day, both makes of fire-fighting apparatus, filling two distinct needs, are manufactured side by side.

Seagrave has offered the collector a fine selection of rugged, powerful trucks to choose from and it is no wonder the marque is one of the favorites in the hobby as well as with the fire departments who ordered the trucks when they were new. I recently had the opportunity to take a trip of several hundred miles in a de-activated 1942 Seagrave V-12 1,000 g.p.m. pumper and can testify that it was a total pleasure during the entire trip (Fig. 4-21).

WARD LA FRANCE

Still operating to this day out of Elmira Heights, New York, Ward La France is virtually in the backyard of its more famous competitor, American La France.

The company continues to be known best for its line of conservatively styled, rugged, custom-chassised fire engines (Fig. 4-22). Since the firm introduced a new line of full-size equipment in 1937, they have been a leader in the industry, although they have never sold the number of trucks as their neighbor with the similar name.

ADDITIONAL COLLECTOR MAKES

John Bean. John Bean manufactured many commercial-chassised trucks for both the military (in 4×4 configuration) and civilian use. John Bean is now a division of FMC Corporation and is operating out of Lansing, Michigan, and Tipton, Indiana.

Bickle. Bickle is one of the few Canadian manufacturers to produce fire engines in great enough quantities that collectors still occasionally run across them. Operating out of

Fig. 4-21. This 1942 Seagrave was purchased by the author's hobby fire company for a modest amount. Although it was missing some loose equipment, it was in prime shape and ran very well.

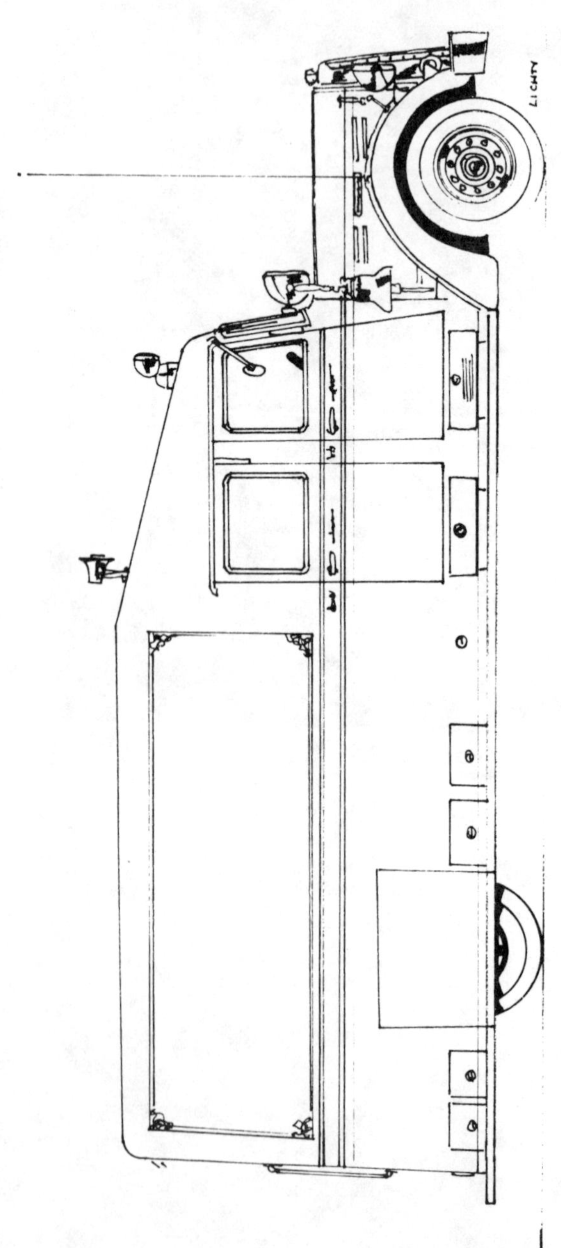

Fig. 4-22. With loads of room and a sedan cab this 1939 Ward LaFrance rescue squad would make a great back-up vehicle for a muster team. Note the unusual skirted rear fenders popular on many Ward La France of this era.

Fig. 4-23. Howe built many commercial-chassised trucks. This early 1950 Diamond T was a very substantial unit and obviously very well equipped for the town of West Lawn.

Fig. 4-24. Stutz was a major manufacturer of fire apparatus. This unrestored pumper from the John Kress collection reveals what an imposing truck the Stutz was.

Woodstock, Ontario, the firm later became Bickle-Seagrave, turning out Canadian versions of the famous Seagrave fire trucks. There is very little difference except the insignia.

Christie. Christie was one of the earliest manufacturers to offer two-wheel front-drive tractors to convert horse-drawn aerial ladders and steam pumpers to the motor age. Operating out of Hoboken, New Jersey, the firm was officially known as Front Drive Motor Company.

GMC. GMC was one of the most important major suppliers of commercial chassis to the custom fire apparatus body builders. Chosen frequently over its sister line of trucks from Chevrolet, General Motors trucks for many years carried heavier, thus preferred, components. Even in later years as General Motors and Chevrolet trucks became almost identical in form and function, the GMC's past reputation as a maker of heavy duty vehicles still caused many fire departments to specify GMC when ordering a chassis from General Motors.

Fig. 4-25. White also built its own fire trucks and offered commercial chassis to other apparatus builders. This is a beautiful example of a 1919 White 500 g.p.m. pumper from Bountiful, Utah.

Fig. 4-26. This Dennis aerial ladder truck is typical of the European rear platform type ladder units. Note how low slung the vehicle is, similar to the low-riding double decker buses familiar in London.

Hahn. Hahn Motors, Inc., in Hamberg, Pennsylvania, is still in business building beautiful full-size custom fire-fighting apparatus as well as commercial chassised trucks. Hahn was known for its rather unusual styling treatments, especially grills, during the early years.

Howe. Howe has turned out to be one of the major commercial chassis builders in the fire apparatus business and it is still operating out of its Anderson, Indiana, home (Fig. 4-23). Howe was one of the first companies to get into the fire apparatus business by outfitting Model T fire trucks. The most famous line of Howe equipment is the *Howe Defender series* of trucks. Common applications are on Ford, Chevrolet, International and Studebaker chassis.

Knox. Knox Automobile Company built cars and fire apparatus out of their Springfield, Massachusetts, factory for many years after the turn of the century. In the days after the horse was retired from active fire service, the Knox-Martin

Fig. 4-27. Federal was a major truck manufacturer through World War II. This 1914 chemical truck is owned by Littleton, Colorado, and still rides on hard rubber tires.

Fig. 4-28. Willy Jeeps were, and still are, adapted to many special purpose type vehicles such as this front mount turret wagon. Four-wheel drive and compact size make these vehicles extremely popular for close-in fire fighting.

Fig. 4-29. Some fire departments even build their own trucks. This long example of a Milwaukee Fire Department-built ladder truck resides in the John Kress collection. The chassis for the truck is also Wisconsin-built: FWD. Note the MFD radiator emblem.

Fig. 4-30. The General/Packard hybrids were not the only Packard-based fire vehicles. This Packard fire engine is from the teens and is shown here during restoration.

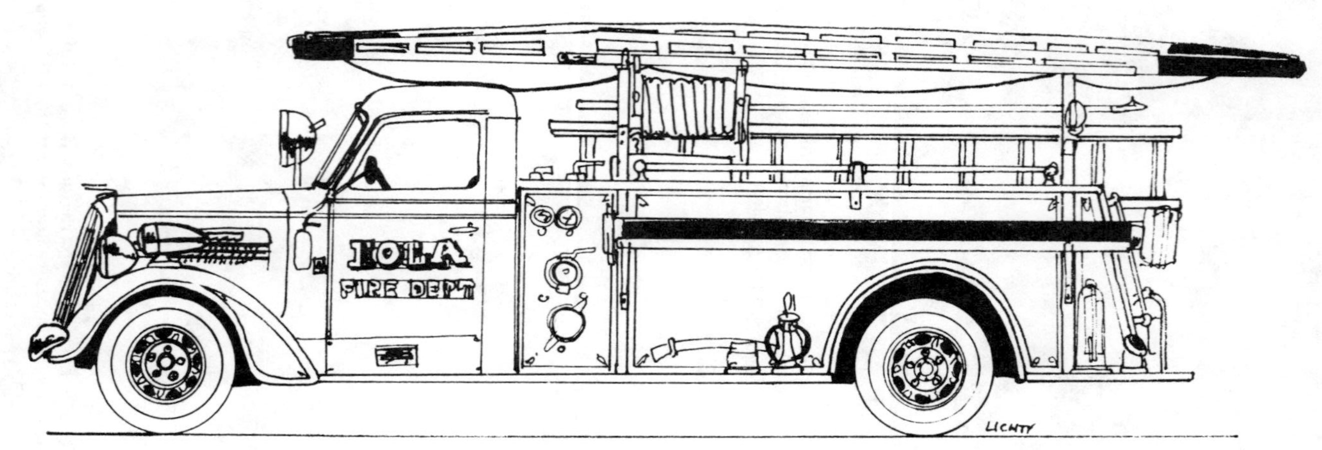

Fig. 4-31. Reo was major chassis manufacturer. This former Iola, Wisconsin, truck has a body and equipment by Anderson and now resides in the collection of Chester L. Krause.

Fig. 4-32. Studebaker made a modest number of chassis for fire apparatus use, but that is not to say the units were not attractive, as this early 1950s pumper clearly shows. Note the smooth lines from cab to truck body.

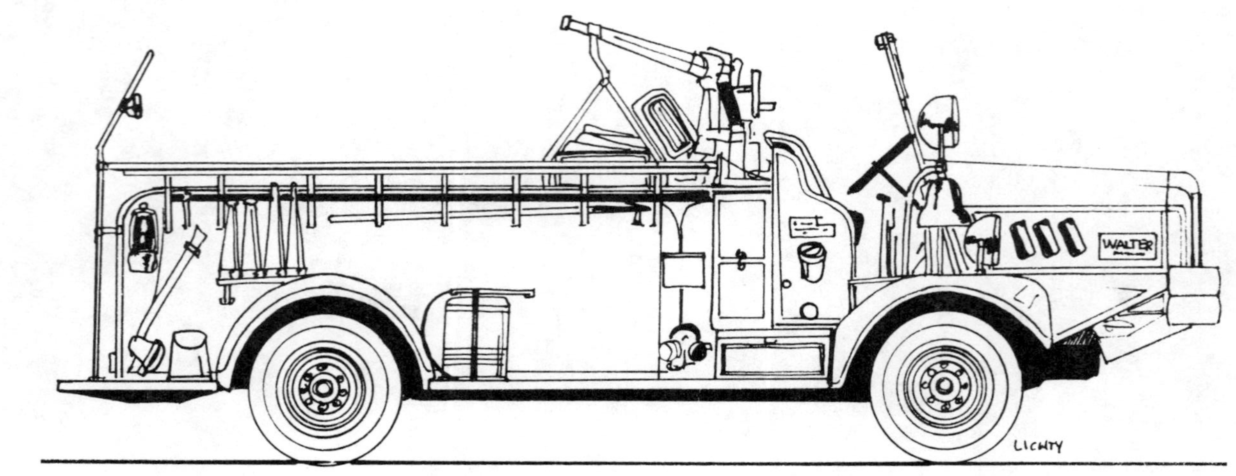

Fig. 4-33. Lide FWD and Oshkosh, Walter is better know for its line of 4×4 snow plows and construction trucks. However, the rugged 4×4 Walter chassis was the basis for many foul weather fire trucks. Walter is especially known for airport crash trucks. This photograph is of a 1936 Walter hose and turret wagon.

Fig. 4-34. Waterous fire engines are among the oldest in the industry. This early 1920 Waterous chemical/hose/pumper is a prime example. Waterous discontinued truck manufacturing to specialize in making pumps. To this day they continue to supply pumps to many apparatus manufacturers.

three-wheeled tractors for converting horse-drawn vehicles have to rank among the more unusual fire-fighting products. It is not uncommon to find Knox equipment on the East Coast, although it is relatively rare as you travel further west of the Mississippi River.

Stutz. Stutz Fire Engine Company (Fig. 4-24) of Indianapolis was also well known for its line of early automobiles, especially the sporty Stutz Bearcat. The firm's trucks carried the same fine reputation and are today highly prized by collectors. The Indy-based firm should not be confused with its Hoosier-state neighbor in Hartford, The New Stutz company which was introduced in 1939, well after the demise of the original Stutz. The New Stutz did carry an attractive line of trucks which would also be good collectibles today.

White. White Motor Company of Cleveland manufactured fire apparatus as early as the pre-teens and later went on to become a major commercial chassis supplier through White division as well as its subsidiary, Autocar division for heavier vehicles (Fig. 4-25).

Others. Other commercial chassis manufacturers (Figs. 4-26 through 4-34) one should be on the lookout for include: Kelly-Springfield, with its unique sloping front end; Cadillac and Kissel, both well-known prestige car builders who offered chassis to early builders; Dodge, Reo and Studebaker, who offered high quality light and medium duty chassis; and, Kenworth and Diamond T, suppliers of heavy duty units to the fire equipment body builders.

You can rest assured that almost every brand of truck that was ever built was at one time or another outfitted with a fire-fighting rig. In your search for a collectible fire engine you may as easily run across a mid-thirties Airflow Dodge as a standard Ford or American La France.

5

Restoration

Before you purchase that special antique fire truck and begin the long, hard battle to bring it back to life, you should first try to determine what you wish to accomplish with the truck. Nobody but you can make this decision for it ultimately determines the amount of time and money you will expend in the restoration. It is therefore doubly wise to make such an important decision before you actually buy your vintage fire engine because the degree to which you are determined to restore your truck should be a major factor in your final decision as to what type of fire apparatus to buy and in what condition.

If you have decided to build a show truck, a fire engine that will win trophies at every concours and car show you attend, you should apprise yourself of exactly what it takes to build such a blue-ribbon vehicle.

It is common knowledge among experienced antique restorers that in the 100-point judging system found most frequently at car shows, the costs of restoring any antique vehicle to where it will consistently win 99 or 100 (Fig. 5-1) points are about double the costs of restoration for a vehicle that will routinely place in the 89-90 range. Depending on who is sponsoring the car show or concours d'elegance, the points a vehicle receives in the judging are based in varying degrees

Fig. 5-1. This beautiful American La France chemical truck from the Harrah collection was as close to 100 points as possible when it left Harrah's restoration shops. Continued use has worn the front tires. The truck was always a popular sight at the annual swap meet sponsored by the museum.

on the common criteria of "fit, finish, authenticity, function, road-worthiness and overall appearance."

When going for those last 10 points that mean the difference between a first place finish and none at all, your fire engine is going to have to impress knowledgeable judges on all of the above points. You will have to plate parts in nickel, rather than chrome, where correct. On an older truck you will probably have to return it to solid rubber tires and possibly even build new wheels on which to mount them because the wheels on older trucks were often cut down to accommodate pneumatic tires later in their careers. Your truck will have to be fitted with the correct number and type of emergency lights and fire-fighting equipment which it carried when new, not with what it may have been decked out with after many years in a fire company. Paint will have to be the correct color of red, not just any leftover barn paint. Upholstery will have to be leather, if that was original, not a Naughahyde replacement. Photos of the truck when it was new will have to be tracked down so that you can reproduce the lettering, striping

Fig. 5-2. Details such as this row of helmets, lanterns and suction hose make the restoration of this little GMC outstanding.

and scroll work accurately. It is easy to see why striving for that last 10 points to perfection can double the cost of a basic restoration.

Only you can decide whether the time and money involved in this exacting restoration will be well spent. A mantle full of trophies is only part of the reward for such care. What may also be important to you is the challenge of restoring a veteran fire engine to factory-new condition. Actually it will be better than factory-new since it is an oft-repeated criticism of current auto show judging standards that an antique vehicle rolling right off the assembly line onto the judging green would never take home a trophy. For some fire buffs this quest for historical perfection and the quiet pride of ownership of a vintage fire engine make all the expenditures of restoration well worthwhile (Figs. 5-2 and 5-3).

For it is a fact that to the novice or the general public a truck that has simply been refurbished to look new, although it may carry some non-authentic or retrofitted parts, is undiscernible from the 100-point show winning entry. The average person does not know the difference between pneumatic tires and hard rubber, or whether your truck was originally built with a nickel- or chrome-plated radiator shell.

If you are not into trophy hunting but still enjoy showing your fire truck at local and area antique auto shows, a 90-point restoration may be all you need or want. Your vehicle will look like new, will function well and will be correct enough to properly impress your family, friends and neighbors. You will feel better about driving your truck across town or to the next county on a Sunday afternoon to participate in a parade or car show without worrying about whether some judge's white-glove inspection of your rig will find bugs in the radiator. As any veteran of the car show circuit can tell you, maintaining a 100-point restoration is also at least twice as costly as keeping a truck in solid 90-point shape. Cost must be another of your considerations in figuring your hobby budget. If after buying the fire engine of your dreams and restoring it to 100-point perfection you cannot afford the time or money to maintain that pinnacle, you will probably wish you had opted instead for the 90-point job that allows you time to take your truck out to shows and some leftover cash to buy a can of brass polish now and then (Fig. 5-4).

There is at least one other level of restoration that deserves notice at this point in your deliberations. That is the restoration of an antique fire engine to a true "fun" truck. Some hobby purists would totally object to such treatment of an historic and storied piece of fire-fighting apparatus, but it might be just the right way for you to go.

I'm not talking about some sort of hot-rodded or show-quality custom project with a jet engine or foot-deep metal-flake paint, but rather the restoration of a fire engine to the point where it may still look factory-new but be outfitted with modern power, convenience and safety options.

If you've bought a fire engine to drive in parades all over the state or to enjoy on quiet country roads every Saturday afternoon, you might well opt for a modern engine that will give the truck plenty of power for today's freeways and a comfortable degree of reliability far from home (Fig. 5-5).

Similarly, if you're a family man who wants to take the kids out for a ride occasionally in the back of your vehicle, you'll probably want to install seat belts, grab bars and other

Fig. 5-3. The same GMC, showing attention to detail such as the dash board, instruments and steering wheel are what separates the extra sharp trucks from the beaters.

Fig. 5-4. This small American LaFrance pumper is a good example of a plain ol' fun truck, with very little equipment and with promotional lettering for a restaurant on the side. It still brings big smiles to the owner's face.

Fig. 5-5. Any custom-chassissed truck like this '20s-era Seagrave will be a better investment than a commercial-chassised truck.

Fig. 5-6. A rough truck like this early Hale pumper will present a million problems, but when tackled will yield the same number of rewards.

safety devices to insure their protection and your peace of mind.

Or, for reasons of convenience of care and cost, you could understandably want to reupholster your truck's seats in something other than the original leather.

For comfort, an add-on air conditioner might be nice, or a stereo tape deck. The options are endless. It's your truck so only you can decide if and to what degree you will blend original equipment and modern options in its restoration.

It is not the goal of this book to tell you which type of restoration is right, but only to inform the prospective buyer of his choices and urge him to set those important goals in advance of actually laying down the cash for such a momentuous project.

WHAT TO LOOK FOR WHEN BUYING

As just mentioned, when you go to look for your first fire engine, you should have already made the decision as to whether you have enough time and money to buy a rough truck (Figs. 5-6 through 5-9) and do a full restoration on it. If that is your desire, decide whether you can handle such a project on a large custom-built truck like an American La France or whether it would be best to get your feet wet with a smaller commercial-chassised vehicle like a Ford or Dodge.

If you don't have the time or skills for such a restoration, maybe you should consider the purchase of a fire engine on which the restoration has already been completed or one which is in fine enough condition to need little or no restoration. While it is likely that your financial outlay will be greater when buying a completed truck, your enjoyment of its use can begin immediately and in many cases, you will be buying another man's time in the form of his restoration labor at a very low rate. Seldom will the value of a restored truck ever cover the amount of time that has been put into its restoration. This goes for you, as well, if you are considering a restoration job on your own truck. Don't expect to be compensated financially for your time if and when the time comes to sell your rig. You'll be putting in hundreds or thousands of hours of work at about a rate of 2¢ an hour. But as mentioned earlier, if financial gain is your principal motive for joining the ranks of antique fire engine enthusiasts, you're in the wrong game anyway.

Fig. 5-7. Closer examination of the Hale reveals rotten hose, weather-checked hard rubber tires and even collapsed sheet metal.

Fig. 5-8. Front view of Hale shows shredded fan belt, corroded metal and bent headlamp/fender support bar.

ROUGH VERSUS CLEAN ORIGINAL

There is nothing wrong with tackling a rough truck for an introduction to the world of antique fire engine restoration and ownership. Just be well aware of the individual truck's shortcomings before committing to such a project. Look the truck over carefully (Fig. 5-10), preferably after finding and studying photos of a similar truck, so that you will recognize any conversions, alterations, additions or stripping that might have been done to the truck in the course of its career or years in collectors' hands. Updating vintage fire engines while they are in service was, and still is, a common practice. These modifications are usually professionally done by the fire department's own shop or, in some cases, the manufacturer of the fire engine. It is not unusual even today to find an old Seagrave in the corner of a custom fire equipment builder's factory having its tired-out V-12 replaced with a new Detroit diesel or receiving other updating.

An important reference tool for the collector and restorer is a book by Walter McCall entitled *American Fire Engines Since 1900*. It was published in 1976 by Crestline Publishing. This book gives a photographic history of the fire engine in the United States that is unsurpassed. The collector or restorer will find himself using this book time and time again to identify trucks or research marques. The book can be even more valuable to the fire engine buff before he makes that first big purchase. The complete range of fire truck photos in the book allows the potential buyer to match a picture with a description he might find in an advertisement, giving him a much better idea of whether the advertised truck may be right for him. The book is available from Classic Motorbooks, Box 1, Osceola, WI 54020, for $24.95, and will be one of the best investments a fire engine restorer will make.

If you are lucky enough to find a clean original truck bought perhaps directly from a fire department, your restoration will naturally be a lot easier. Hopefully, the truck will even contain most of its original equipment.

SOURCES AND METHODS FOR BUYING

One of the best sources for buying a collectible fire engine is from its original owner—the fire department (Fig. 5-11). It would be a most callous fire department, indeed, that would kick a veteran truck out of the garage when it becomes

Fig. 5-9. You can bet the pump on the Hale is as corroded on the inside as it is on the outside. Complete rebuilding will be in order.

Fig. 5-10. Look closely at the engine on the Hale. Bent and destroyed parts will have to be replaced, possibly even made by the restorer. Also, there is a part missing from that hole on the side of the block. If the restorer doesn't know what it is, he's going to need a shop and a parts manual.

obsolete with no consideration of its historic past. Yet, tight municipal budgets and that familiar old problem of all fire engine enthusiasts—lack of adequate storage facilities—usually prevents the department from keeping a truck when it is put out to pasture. It is not uncommon, however, for a fireman or a group of smoke-eaters to buy the truck and at their own expense restore, maintain and muster a favorite vehicle. The pull of nostalgia that drives many non-professionals to the purchase of an antique fire engine works even stronger on the fireman who may remember the many hours he spent polishing the brass or tinkering with the engine on an old truck during his rookie year with the department many years ago.

The first stop for the prospective fire engine collector should logically be his local department. Introduce yourself to the chief or assistant chief. He will usually be pleased to hear of your interest in collectible fire apparatus and in his own department and its equipment. He may or may not know of imminent plans for the retirement of one of the trucks in his care, but will often be an excellent source of leads for you at a

Fig. 5-11. This late 1940 American LaFrance aerial ladder was put out for bids by the Wausau, Wisconsin, Fire Department several years ago. As you can see, the hobbyist bidding would have to do virtually nothing to be ready to enjoy this rig. Naturally, a collector can afford to bid high on such a nice truck, especially if it still carries most of its equipment.

fire department in a neighboring city. While you are at the firehouse, ask to take a look at the chief's current equipment. Usually he will be flattered and you will get the opportunity to see fire engines as they are really used, along with the quantity and type of equipment they carry while in active service.

Don't be afraid to get around to the surrounding communities either. Visit with the chiefs and ask to see their trucks, especially any that might be nearing replacement. Be sure to ask the fire chief if he subscribes to a state fireman's association magazine or perhaps a special publication for fire chiefs in your state. These magazines frequently have ads for trucks that are available and are a good possible source for the collector to pick up a newly-retired veteran vehicle.

When a fire engine is sold, however, it is seldom just handed over to the first buyer who walks into the station with a cash offer. Because the department is usually an arm of the municipal government and the trucks are the property of that government and the citizens as a whole, it will most likely be required that the department offer an obsoleted truck in a public bid. In most cases, ads soliciting sealed bids for the truck are placed in local newspapers, firemen's magazines and perhaps even in the antique auto publications. On the specified day, the bids are opened and the highest bidder owns the truck. Other departments will announce the sale of a truck and invite buyers to attend a public auction. You have to be pretty sharp to spot these ads because they are usually small and may be tucked away in the Public Notice section of the classifieds among announcements of city council committee meetings and weed control warnings. Be sure to keep in touch with the chief and you will be less likely to miss the opportunity.

There are not a lot of secrets to bidding on a fire engine, but a few basic rules should be kept in mind to increase your chances for success. Surprisingly, your competition for virtually every obsoleted fire engine that comes up for sale will include the local scrap dealers whose only purpose in obtaining the truck is to cut it up and crush it down. Therefore, the first calculation you will have to make if you are going to cast the winning bid will be how high the junk dealers can go for the truck and still make a profit. To do that, you'll need to know as exactly as possible how much the truck weighs and what the current rate for scrap metal is in your area. When figuring this scrap value, don't forget the truck may contain a significant quantity of brass, copper or nickel which help drive the junk value above what it would be for just scrap steel. With accurate calculations you will be able to determine where the junk dealers will have to draw a line. Remember, they have to be able to put a great deal of labor into turning a fire engine into blocks of crushed steel. Therefore, it may be easy to save the truck you want from the crusher!

Next you will need to figure the cost of restoration as it applies to your hobby budget. You should have already decided the degree to which you will restore and maintain your fire engine, so look the offered truck over and attempt to find out what it will need in the area of restoration compared to the standards you have set. Offer your bid accordingly. Do not let yourself in over your head in relation to what the truck's value will be when restoration has been completed (Figs. 5-12 through 5-21).

Another guideline which may aid you when setting your bid will be the value of similar trucks in the collectors' market. For this, you'll want to check back issues of fire vehicle restorers club publications and the commercial antique auto papers. However, remember that the advertised selling prices are seldom realized and should be used only as a general market guide.

You might also win a fire engine bidding competition with just a bit of individualist thought. People tend to value such items as antique motor vehicles in round figures such as $1,000, $1,500 or $2,000. Try bidding an odd amount just over such logical prices such as $1,025, $1,505 or $2,002. That extra couple of dollars might be just the right amount to top the second highest bidder and get you the truck you want. If should also be comforting to know that at least one other person thought the truck you bought was worth almost exactly what you had to pay.

But don't feel too bad if you outbid the second highest

Fig. 5-12. Detail items like this rear searchlight on a 1927 Seagrave add to the vehicle's desirability when you are purchasing it. It is one less authentic part you will have to find and pay for later.

bidder by a wide margin to win your fire engine. That only means you wanted that particular truck more than anyone else who knew that it was available. This is especially true if the

Fig. 5-13. On the same Seagrave, examine the dash. The scaley paint is unimportant because it will be repainted anyway. Check for complete gauges and lights and for broken castings.

sale of the truck was poorly publicized. If you have bid an amount you were comfortable with and won the truck you wanted, you should have no regrets.

By the same token, don't feel bad if you are widely outbid. It only means that the buyer had a better use for the truck, or simply wanted it more than you did. The bid system is fair for buyer, seller and the public. It assures that the department gets top dollar for its vehicle in a competitive atmosphere and that the mayor's nephew or a scrap dealer promising a kick-back aren't allowed to steal the truck for a song. For the fire engine collector, the best part of the bid system is that you have the chance to offer the seller exactly what you want to pay for that particular truck. Thus, you either buy the truck at your own price or you don't get it. By setting that price yourself, you should never feel that you were taken advantage of in a deal.

Another avenue to take in your search for the fire engine of your dreams is to write letters to communities that you know have owned trucks of the type you desire. Just because

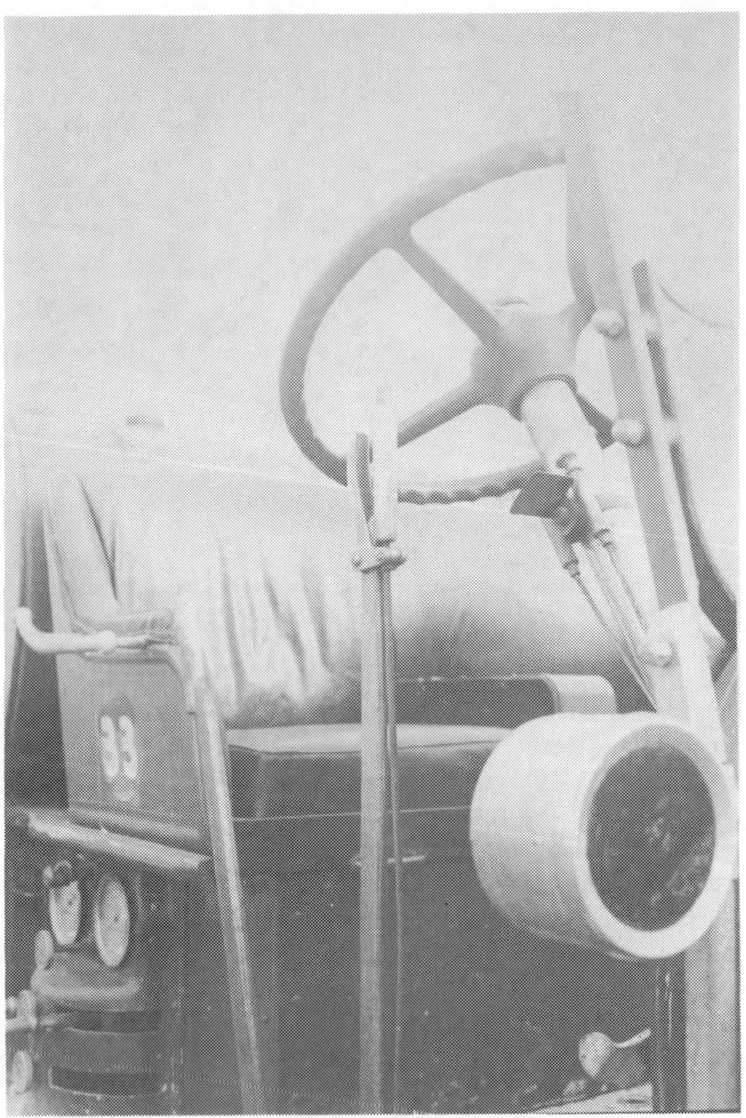

Fig. 5-14. Fortunately, the upholstery on this Seagrave is almost intact, but savings there will almost certainly be spent on plating. Look at the hand brake and shift lever. The red cowl lamp is a valuable extra.

you don't see the truck you remember that department once used does not mean it has been sold. It may simply have been relegated to a reserve status in a dusty corner of a city garage.

Fig. 5-15. The engine compartment of this late 1930 General sedan cab pumper looks intact and complete, but will the truck run? It is not too much to ask the owner to produce some batteries and give it a trial cranking over.

Fig. 5-16. This 1948 closed cab American La France was once a spectacular pumper. Unfortunately, the former owner robbed it of its entire rear body and pump. Even in prime condition and with all of the lights, replacement of the body is such a major project that it makes the truck almost worthless. When figuring a price for this truck, a bidder would have to consider a complete new body for the rear. Tracking down a used one will not be easy and it is unlikely an exact replacement will be found.

Fig. 5-17. The dashboard of this 1936 Ford open cab pumper looks good superficially, but a closer look shows holes have been made for auxiliary gauges, there is a non-stock ignition switch and you will have to find an original replacement for the late 1950s-style steering wheel.

It is not uncommon for a department to put such a truck away for emergencies that never happen and then forget completely that they still own it, until a collector comes along and reminds them. If you can ferret out a remembered truck in that fashion, you might just cause it to be put up for bid.

The next important step in your search for an antique fire engine restoration project should be to join SPAAMFAA—The Society for the Preservation and Appreciation of Antique Motor Fire Apparatus in America. Currently, the dues are $10 a year. These dues include a subscription to the Society's excellent quarterly magazine, a periodical that features marque histories, how-to articles and, best of all, advertisements offering fire trucks and equipment for sale. The club and its chapters are also active in vintage fire equipment musters. They sponsor many of these events all over the country each year. Joining SPAAMFAA is a must for the serious fire engine collector or restorer. The address is P.O. Box 450 Eastwood Station, Syracuse, NY 13206.

The various antique automobile collectors' periodicals

Fig. 5-18. As mentioned previously, the best place to buy a fire engine is from the original owner. This FWD still belongs to Bear Creek, Wisconsin.

Fig. 5-19. Antique auto swap meets like the giant one at Hershey, Pennsylvania, can produce finds like this early 1950 Ahrens-Fox pumper. It is surrounded by parts vendors and debris after a week's worth of swap meeters rummaging around it.

are also a good source for advertisements of trucks for sale. If you are just beginning your search for a fire engine, you might want to try a subscription to *Old Cars*, a weekly tabloid newspaper. That paper offers a special column for fire engines in its classified ad section. Your $13.50 annual subscription will buy you 52 opportunities to look for a collectible fire truck. After you've found your truck, you can also take advantage of the many articles, features and columns the paper offers. In addition to how-to columns and features, *Old Cars* also has a regular writer specializing in antique trucks who often touches upon vintage fire engines. *Old Cars* can be subscribed to by writing to 700 E. State St., Iola, WI 54945.

A monthly publication that contains solely collector vehicle ads is *Hemmings Motor News*. It looks like a small phone book and frequently offers many fire engines in each issue. Scanning current copies or recent back issues should be useful when trying to get a feel for the price structure of the vintage fire engine market as a whole and in particular for the truck you are considering. Available by the single copy at

Fig. 5-20. Antique auto actions and car corrals are also a good source for fire trucks. This 1936 Ford was found in a car carral sales area at the Indianapolis-based Hoosier Auto Show. That '53 DeSoto wouldn't make a bad fire chief's car, either.

many larger newsstands around the country, *Hemmings* can also be ordered on an annual subscription basis for $8.75, from Box 380, Bennington, VT 05201.

While on the subject of buying fire engines from collector car publications, it is interesting to note that the average antique auto enthusiast with no special interest or expertise in the field of fire equipment usually nutures the misconception that the value of any antique fire engine he may have or stumble onto should be in direct proportion to the physical size of the truck. While this rule of thumb is probably true when dealing with Packards or Cadillacs of the same eras, this couldn't be farther from the truth where fire engines are concerned. However, it is good for the fire buff to be aware of this attitude and beware of inflated prices when dealing with a car collector.

The last, but not necessarily the least important sources of a collectible fire engine are the manufacturers and dealers of such equipment. It is not uncommon for them to take a trade-in on a new truck, although the typical trade-in is a truck of a newer design that can be used by a smaller department or refurbished and put back into active service. It is unlikely that you will be interested in spending the type of money a factory trade might cost because you will be competing in a market where the trucks usually retain relatively high utility as emergency vehicles and thus, higher values. With a new aerial ladder truck selling in the $250,000 range, don't be surprised to see a reconditioned used specimen go for $50,000; or, for that matter, a junker sell for $5,000. Don't be disheartened by the fact that you probably won't be able to buy a collectible truck right from the factory though because there is often a dealer or factory representative around who appreciates what you are looking for and he might provide you with an important lead.

RESEARCH THE VEHICLE

Once you have obtained your antique fire engine, the next important step is to research the vehicle. Most fire engine enthusiasts want to find out all there is to know about the unique history of the truck they have purchased.

Fig. 5-21. Antique auto shows can also produce some good buys. This beautifully restored Reo/Boyer chemical truck was being offered by an owner anxious to sell due to a job relocation.

Start by contacting the manufacturer. It is common in the industry for the builder to retain factory work orders, blueprints, work book and even photos of your truck before it was delivered. This is the single most valuable source you are likely to encounter in your search for information about your vehicle.

It is likely that if the factory still had its information on the building of your truck, they can also tell you to what fire department it was originally delivered. You will want to contact that department as the next step in researching your truck. Ask for photos, histories and anything else they might have relating to your truck. They might even have a piece of equipment the truck had been outfitted with during its career. If the department that was the original buyer sold the truck to another fire department, re-trace your steps. If it was sold to an individual, contact him. Any of these sources might still have original equipment you can purchase. It is easy to remove a ladder from a truck while it is in storage and then forget to put it back on the truck when it is sold.

Also consider the contemporary accounts in which your truck might have figured. Contact local newspapers where your truck served, or the town's historical society. They might have photos of your truck battling great fires in the town's past. They might even be able to furnish records of the truck's purchase by the department, or its subsequent sale.

Check your membership directory to SPAAMFAA. See if there are other members with trucks similar to yours. Contact them and keep in touch. Remember that you share a mutual interest. Your fire engine can lead to many lasting friendships over the years.

Another source of information about trucks like yours would be the various fire and antique auto museums across the country. A contact with a curator or manager may produce useful information about your truck. You can never research your truck too well. Every bit of information you glean will make your restoration easier and your pride of ownership greater.

Final sources of both research information and a list of fire engines for sale are the various trade or society journals

(Fig. 5-22) of the fire-fighting and fire-equipment professions.

These are aimed at the fireman, but many have ads for retired apparatus as well. Most carry interesting articles and photos of early fire engines. These professional journals include:

- *International Fire Chief,* 1329 18th St. NW, Washington, DC. Subscription information not listed.
- *Firehouse*, 515 Madison Ave., New York, NY 10022. Annual subscription price, $15.
- *Western Fire Journal*, 9072 E. Artesia Blvd., Suite 7, Bellflower, CA 90706. Subscription information not listed.
- *Fire Command*, 470 Atlantic, Boston, MA 02210; $10 per year.
- *Wisconsin Fire Journal*, 4289 Beltline Highway, Madison, WI 53711; $6 per year.

WORK AND STORAGE FACILITIES

Where and how you will work on the restoration of your new-found prize will usually be determined by your existing facilities. Few of us have the financial capacity to build a new

Fig. 5-22. Publications will be invaluable to you throughout your restoration project.

building every time the old one does not fit what we are doing (Fig. 5-23).

The first rule is to measure the truck *before* you buy it, not after. Also measure your garage. It is as simple as that—either it will fit inside your garage or it won't. Don't try to squeeze a truck in too close because you will want room on all sides of the truck to work during the restoration. Cramped quarters can be an especially severe handicap when restoring a fire truck that will involve moving large, heavy parts around.

When you measure the truck, be sure to include lights, sirens, or whatever else may protrude above the top of the cab or windshield. The extra height of a fire engine, as often as its longer-than-average length, can create storage problems for the owner. The typical garage door is seven feet high, and very few fire trucks will fit through it. If you have no alternative facilities available, you may have to consider the purchase of a new garage door and plan a couple of weekends to work enlarging the opening before your restoration can commence.

The ideal workshop for this type of restoration project will have extra room to work in (Figs. 5-24 through 5-26). A two-stall garage would be just fine. By having this much room, major pieces can easily be removed and worked on near the truck on the open floor space. This will also provide room for the adequate use of chain hoists swinging around carrying pumps and engines. Even a simple part on a wheel and tire for a fire engine can weigh several hundred pounds. They will also require some special equipment and effort to move unless the restorer is a weight lifter.

Be sure to have a good sturdy work bench and plenty of shelving. It will facilitate the storage of your parts in a systematic order. For screws, nut, bolts and other hardware, use glass jars. You'll need lots of them in sizes from baby food jars to quart mason jars. Many restorers find it convenient to nail the lids of the jars to a long board suspended from a cabinet or other high member. Then the jars are always in order, can easily be screwed on and off and are kept off of the counter, freeing that space for more important work. A series of sturdy corrugated cardboard boxes in various sizes and shapes will also be handy for the slightly larger pieces.

Don't trust your memory. Be sure to label everything and label it with something that is waterproof and will not fade. Not all felt tip pens provide permanent marking so select your writing utensils carefully. It's also a good idea to make your labels bold. Perhaps you could even color-code them so that a few greasy thumbprints won't obscure your information.

BASIC DISASSEMBLY

Once you have your garage set up the way you think it will be most efficient for your restoration, you will be ready to

Fig. 5-23. This old camper showroom makes an ideal storage place for this Pierce-Arrow/General pumper.

Fig. 5-24. A work area is even better with nearby work benches and plenty of light. Special features of this truck include an eagle radiator cap, a ratchet-type crank starter and an odd-shaped high pressure ball.

start your project. Kiss your wife goodbye, tell her you will be out in the garage until further notice and assure her that it is better than spending an equal amount of time and money at the corner tavern. If you can arrange it beforehand, it wouldn't be a bad idea to take her for a ride in a friend's fire engine to reassure her that it will all be worth it. Of course just one ride in a fire truck for your kids will win them over to your side immediately!

Since it is unlikely that you will have a high pressure water jet cleaning system at home, your first step if the truck is running should be to take it to a local high pressure do-it-yourself car wash.

Arm yourself with a large quantity of Gunk or other such de-greasing agent and possibly a long-handled brush for hard-to-reach areas. Wear high rubber boots and a rain coat. Be sure to take plenty of quarters because it is very disconcerting to get halfway through this initial cleaning process and run out of money to feed the car wash.

It will be necessary to cover water-sensitive components such as the carburetor and distributor with heavy plastic while you are cleaning the truck. With these important

Fig. 5-25. This late 1940 Mack has dark, cramped storage, but at least it's indoors and dry.

parts covered, you don't have to be afraid to clean the chassis and underside where you will find caked-on grease and grime. Remember, the cleaner the truck is to begin with, the easier your job will be during restoration.

Once you get the truck home, let it dry out thoroughly and clean it with a whisk broom and vacuum cleaner to get rid of dust, dirt, dried leaves, etc. The less particles you have falling in your eyes as you work, the better.

There are two schools of thought in restoration circles about refurbishing parts. One school holds that you should restore each part as it comes off the vehicle because you have just seen how the part was fitted, where it went and exactly what you need to do to it.

I favor the other school of thought: disassemble the truck and parts first, then restore them as you put it back together. Restoration of an antique vehicle on the scale of a fire engine can take a long time and freshly restored parts can age, even just by sitting on a shelf. Paint and plating can become discolored or even shelf-worn from being banged around. By the time the truck is ready to be put back together, you may find you have to *re*-restore many pieces.

As a compromise to these two schools, it is a good idea to send off any parts that may need such specialists as platers and rebuilders as soon as they are removed from the vehicle. For the most part they are very slow and may finish your job just about the time you really need the parts for reassembly. Be sure to take an accurate inventory of any parts sent away. It is even a good practice to take a photo of the parts layed out on a clean table before you send them to anyone.

As long as you have your camera out and before diassembly, take several rolls of film. Shoot photos of the truck and its components from every angle. Shoot into every opening. Take photos from underneath and on top. Open each compartment. Even if it appears to be a simple part, take a photo of it. Shoot the engine, transmission and pump assembly from all sides. Parts that appear identical may vary slightly in reality or in how they are attached to the vehicle. You can't take too many pictures. This photo album will literally act as your shop manual later in the restoration.

Fig. 5-26. While not a fire engine, this Sterling truck is a prime example of what prolonged outside storage can do to a vehicle.

Of course an actual shop manual, if you can obtain one, would be an immense help. Even original blue prints and factory work books help show details you may have missed on the actual vehicle. Obtain as many of these types of books as you can when you are doing your pre-restoration research.

As you begin removing the parts, take more photos. Take photos after each major step. Start a diary and log everything you do during the disassembly. Some restorers even use a tape recorder and make comments throughout the restoration. Some of these tapes can be quite humorous when comments have been recorded immediately after a wrench slipped and a finger was slammed into an unyielding piece of fire engine.

Again, don't trust your memory. A year later you will not remember every part of your truck, no matter how intimately you feel you are getting to know it.

Identify each part with a paper tag and put it into storage either by itself or with parts of absolutely identical shapes and sizes.

When putting parts away, it is a good idea to further clean any remaining grime or rust from them. You may elect to leave some subassemblies together to restore as one piece later and there is nothing wrong with this. While having no detrimental affect on the quality of your restoration, it may save you time and frustration. Parts of this type might include the generator, distributor, pump, chemical tank, etc.

Try to store the parts in some type of logical order so you can lay your hands on them when you need them. Give some thought to this sequence because you do not want vital parts to be buried when you want to work on them. If this is your first acquaintance with an antique vehicle restoration project, you will be shocked by the sheer number of pieces there are in this large fire engine with complex special equipment.

While the truck is a complex piece of machinery, the tools you will need for the restoration do not have to be. A good basic tool box containing an assortment of open and box end wrenches; large and small socket sets; a good set of screwdrivers; and the universal tool, a large hammer. Keeping in mind the size of the vehicle, you can imagine that the tools involved will have to be substantial. If the only vehicle you have ever worked on has been your imported Japanese family car, you will need to supply yourself with some larger sizes of tools. Throughout the restoration you will find the need for several unique tools which you will have to obtain as you go. If you can find a fellow restorer who has had experience with fire engines, he may be able to advise you on any special items he thinks will be necessary for your particular vehicle.

Be sure to have plenty of lubricant such as WD-40 or some other good penetrating fluid. A *nut splitter* is an excellent tool to have for stubborn nuts that won't come off even when soaked with lubricant. You will also need at least a small propane torch. You may have to heat some pieces to loosen them. You may or may not elect to purchase your own welding outfit. Acetylene is not hard to work with once you learn the basics and there are many fine books showing you how to handle it. Other hobbyists elect to farm the welding and cutting out to a local shop. However, with the size of a fire engine and its parts, this is sometimes impossible. It may prove equally impossible to get a torchman to make house calls.

An air compressor of a modest size may also prove to be a major aid in your project. Even if you do not intend to do the final paint job on your truck (and there is no reason why you should be afraid to tackle it), you will want a compressor for priming, cleaning and possibly for operating an occasional pneumatic tool.

Some hobbyists cooperate and share major tools and equipment. One will buy a welding outfit and another will buy an air compressor. This is not a bad route to take if you get along with the other hobbyist well, are prone to return things and if you will both not be at the same point of your restorations at the same time. Perhaps if you are already a member of a local antique auto club there is such an arrangement among that membership.

While fire engine bodies are basically very simple and very rugged, the years in which your vehicle was in service will no doubt have left their mark and you will want to do some body work in the course of your restoration. For these repairs, you will need a mechanics dolly, or creeper (which you are likely to find useful even if you are not undertaking any sheet metal work on your truck), a basic set of body hammers and perhaps special equipment such as dent pullers, etc. However, before you rush out and buy a complete set of tools that would outfit a body shop, assess your own needs in terms of the work you feel necessary and buy your body tools accordingly.

Details on the art of body work won't be covered in this book since the basic techniques and skills are identical to those used in all other fields of vehicle repair and restoration and the many books available on the topic should be able to teach you all you need to know.

Most fire engines contain wood parts or equipment as well as metal. Older trucks will have wooden wheels, hose beds and perhaps even body framing; while later model trucks may have only a wooden ladder or two. In any case, check to see that you have a few basic woodworking tools such as saws, files, etc., on hand for the restoration.

A large shop-type vacuum cleaner, a broom and dust pan will also come in handy. As you get into your restoration, you might also appreciate having a trash bag dispenser near at hand. If the beams and joists in your garage can support the work—and you'd better be sure they can before you try—you should probably consider investing in a block-and-tackle type chain hoist for lifting and moving major components. While such gear can easily be rented, the frequency with which a fire engine restoration calls for its use could make rental that often or for that length of time a financially unsound proposition. Besides the creeper mentioned earlier, you might want to buy or make a sturdy dolly from heavy duty casters and stout lumber. This dolly will facilitate the movement on the floor of the bigger parts. Items such as this can take much of the back-breaking labor out of your restoration job.

THE MECHANICS

As you disassemble your fire truck, you will eventually

get down to the basic chassis and running gear. You may be surprised to find that two common words describe virtually every fire engine's drive train and related components: massive but simple. Older fire engines are usually powered by T-head 4- or 6-cylinder engines of simple, straightforward design. Later the engines became more complicated and V-12 powerplants were used almost industry wide. Of course any apparatus built on existing commercial chassis such as Ford and Chevy use the respective power plants of those makes.

The restoration of a fire engine's drive train should not present any problems or challenges that would not be similarly encountered in the same type of engine because they were also used in many other vehicles. In fact, the legendary care which fire departments lavish on their rolling stock should mean that your truck's engine will be in much better preservation than it would if it had been used in a moving van or auto. Likewise, due to the emergency nature of the vehicles, they are likely to have very low mileage compared to the miles they would have received in another type of vehicle. Low miles can often help to insure a relatively easy time of restoration.

One common peculiarity of fire truck engines which you should watch for is the twin- and sometimes triple-ignition system that might have been employed. A more complex distributor and magneto system and more spark plugs than you'd see in a Champion warehouse are included in these systems.

Fire truck transmissions range from simple three-speeds to complex 5-speeds with overdrive, power takeoffs to run fire-fighting equipment and sometimes two-speed rear differentials. The basics are the same as any other automotive transmission, however, and should present no special problems. Once again, the size and weight will be your greatest obstacle when dealing with these parts.

To this day, even new fire engines are suspended by leaf springs with an I-beam from the axle and standard rear differential. The only exceptions would be the four-wheel drive types in the many variations in which they appear. You will want to strip your truck right down to this level. Disassemble the springs to check for cracked leaves and to clean them of rust and scale.

Instead of standard lubrications, many restorers are now using teflon tape in springs to eliminate squeaking and further deterioration. Obviously this is a departure from original equipment and it might violate the sensibilities of a concours judge or a fire engine purist, but the teflon tape is not visible after assembly and makes your fire truck a better wearing and better driving vehicle.

Modern products and chemicals can aid the fire engine restorer in virtually every step of his project and where they do not affect the visual authenticity of the vehicle, they should be used to make the job easier for you. So-called *locking chemicals* such as Lock Tite can be used to fit and retain secure parts. Also, the new, hard *Emron* type paints can assure a permanent hard gloss finish that can be enjoyed for years to come. These Imron paints are almost a must on chassis parts and it is worthwhile noting that most new fire trucks are now being painted in this trough, hard paint at the factory. If you are planning to use and enjoy your vehicle, the whole truck could be painted with this material.

FIRE EQUIPMENT

You will find yourself getting into the unique challenges of fire engine restoration right at the chassis level. The previously mentioned power takeoffs are only the beginning, supplying power for such things as pumps, generators, hydraulic systems, etc. (Figs. 5-27 through 5-33). Some are gear-driven, others utilize hydraulics. Really old aerial ladders are spring-powered and always require extra caution by the restorer and user.

As mentioned previously, there are several types of pumps found in the fire trucks of various manufacturers. There are rotary, piston, centrifugal pumps and others. You would be wise to obtain a manual for the type of pump installed on your truck. If you plan to muster your truck you will want to have your pump fully operational. In that case, you'll need to make sure that it pumps to full capacity and is properly

Fig. 5-27. This pressure ball on another ALF pump has no real working parts but must be in top condition and free of defects for proper pumping after your restoration.

sealed against leaks. It is interesting to note that some of the old original component pump manufacturers such as Waterous are still doing business and providing pumps to major fire

Fig. 5-28. This high pressure deluge gun, or turret nozzle, would have to be plated and all of its valves would have to be checked before use in a muster.

Fig. 5-29. The heart of any pumper is the engine itself. This American La France V-12 with twin ignition is not really all that different from its Lycoming/Auburn heritage engines and should pose no special problems to the average shade tree mechanic.

truck builders like Pierce. It is safe to assume that these on-going concerns would be an excellent source for parts, seals, technical specifications and advice when re-building your truck's pump.

Another obstacle not infrequently encountered by the fire engine restorer is the fact that your truck's pump may be missing altogether. Due to the high concentration of brass in its make-up, a faulty or old pump is often scrapped off of the truck before the hobbyist gets his hands on it. That will bring you back to the drawing board once again. Economics will most likely prevent you from purchasing a new pump and that would most likely not be authentic, anyway. So the next step will be to check with your fellow fire engine restorers. You could even place Parts Wanted ads in the hobby publications.

In direct contrast to the complexities of the pumps, aerial ladder systems are simply mechanical systems that anyone with a basic understanding of mechanical principles or even just a good dose of common sense should be able to figure out and rebuild (Fig. 5-33). Most aerial ladder systems are operated on hydraulics, springs and sometimes air. Infrequently, they are even gear driven.

It is not a bad idea before you disassemble your truck to

Fig. 5-30. You may have to develop electrical skills when special equipment like this Mars light gives you problems. It has a gear driven motor to swivel it horizontally. The bell is rope powered and if you can't make that work, you're in over your head. You better sell the truck.

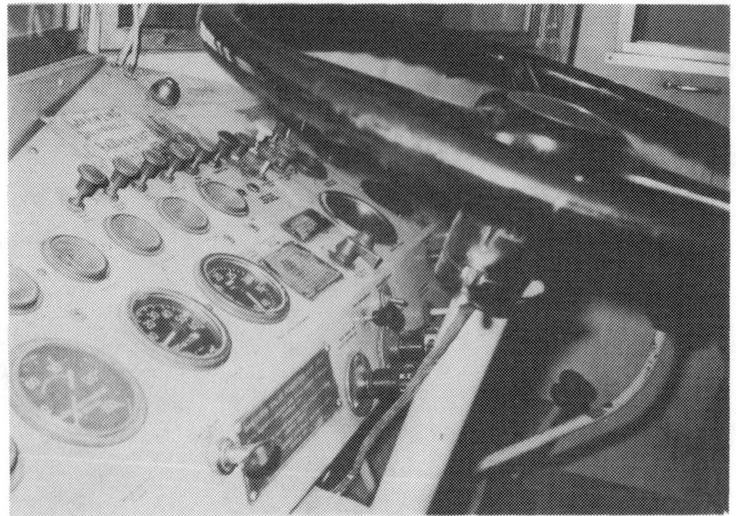

Fig. 5-31. Another electrician's nightmare, this dash board from a Marmon-Harrington/American La France airport crash truck has switches and gauges galore to give readouts and control the vehicle's three powerplants. It would be good to obtain schematics from the manufacturer before tackling any major rewiring here.

have tested many of these fire apparatus assemblies. You will then be able to determine just what needs rebuilding and what does not. There is no sense thinking you've completed your restoration only to find you have to remove a major component for rebuilding. There is never a shortage of volunteers around when you decide to fire up the old pump for a test squirt, especially on a hot summer day. Kids young and old seem to love to get wet, particularly when they can do so while playing with a vintage fire engine.

PAINT

When preparing your truck for paint, and before tackling any body work, you will once again want to click your camera for a complete set of photos of the entire vehicle. If you are lucky enough to have a truck so original that it still possesses its working complement of striping, lettering and scrollwork, be sure to go to an artist's supply store for some large sheets of tracing paper. Trace every last detail and shadow of the fancy painting so you will have a pattern later when you are relettering the vehicle (Fig. 5-34). Also be sure that you tape down the paper when tracing so that it does not slip and distort your image. As a final step to assure yourself of proper perspective and alignment when you repaint these details, mark your tracing paper with the exact location in reference to parts and holes on the body.

Body work on a fire engine is pretty much the same as on any other antique motor vehicle. If you are a novice, there are many well-illustrated, self-explanatory books available to show the amateur how to do basic body, fender and paint restoration. The main thing is to have patience when learning and practicing your new skills. The body and paint end of your restoration project will be the most visible of the entire job

Fig. 5-32. These beautiful hammered chemical tanks from Harrah's American La France have no real complex workings but would be extremely difficult for the average hobbyist to reproduce.

Fig. 5-33. On the restoration of an aerial ladder turntable and lift mechanism, you'll need a basic knowledge of hydraulics and gears. The sheer size of the equipment makes the system fairly easy to understand.

and you should be able to see and enjoy your accomplishments clearly when you are finished.

It is not unusual to find many odd holes in your fire truck's body where new equipment or modifications had been added. You will want to braze all such holes shut and finish them to a point of invisibility. On larger holes you may have to weld a piece of metal into place. A technique known as *hammerwelding* is used for such jobs. It takes some practice as a welder, so you may want to call in the aid of an expert for any really large holes to be filled.

Many early fire engine seats and bodies, not to mention hose racks, contain lots of wood. Most of these parts can be made with simple home woodworking tools and techniques. Almost none of these wooden parts are exposed like on the woodie station wagons (Fig. 5-35), so your level of skill and the finishing of the woodworking will only apply to fit and function. Virtually all of the wood is eventually covered by upholstery or body metal.

Of course any wood that does show must be authentic

Fig. 5-34. Decals may be available to replace the scrollwork on the Seagrave aerial ladder shown, but don't take any chances. You might have to hand paint the designs back on after the truck is repainted. Take lots of photos and make tracings of old designs and lettering.

and once again you might find yourself calling in an expert if you determine that the work is beyond your ability. Even the wood that does not show should be as authentic as possible, as often when it is attached, its shape or strength can be apparent. Refrain from doctoring the existing wood. Most likely it is rotten or dry rotted and any makeshift patchwork will come apart soon after you reassemble it.

Most of the woods used in early vehicle construction were of the dense, hard variety such as oak or a hard rock maple. In areas where the wood does not show, it will not matter too much what variety you use, as long as it is hard, rugged and not prone to rotting. However, where such parts show, it is advisable to use exactly the same kind of wood as the original, especially where the wood is finished in its natural color. Most lumber yards can provide you with the needed woods, or in case of special hardwoods, they can advise you where to obtain them.

PAINTING YOUR FIRE ENGINE

As mentioned earlier, painting an antique fire engine is not perceptibly more difficult than painting any other vehicle. However, there are a few special tips to keep in mind.

One is that the complexity of your truck will dictate that you will have to paint some of your parts off of the truck in order to obtain the finish you desire. To do this you will need to construct a rack or hanging bar in your paint area. Each component can be hung up while painting and you will have a minimum of surface area marked by the hanger. Many painters simply use coat hangers or other stiff wire for this purpose (Fig. 5-36).

Prepare your parts as you would any paint job, working your way up to a 600-grit fine sanding. Make sure you do not get dirt or grease into the paintwork. Make sure all body panels are well degreased before you ever touch them with a body grinder or sandpaper. You may think that sanding will remove the oil or wax from the finish but that is not true. It only serves to grind these elements into the base coat of your new finish, creating a fish eye or pock mark effect in your paint. Even sandblasting a part that is waxed or dirty can

Fig. 5-35. While not many fire engines have their wood body supports exposed to the elements as does this Buick woodie, it is likely that dry rot has been at work on the hidden wood as well. Luckily trees are still growing as a parts source is always available. Just get out your saw. Hardwood and milled lumber is available at most lumberyards. There is no such thing as automotive wood, just good hardwood.

Fig. 5-36. Hanging loose parts up for painting is the way the pros do it. Body doors are shown after having just been painted at Pierce Manufacturing's final assembly area.

further imbed the oily substance into the metal. There are a number of commercial *metal prep* type solvents with which the surfaces can be washed to prevent this. Check your local auto body paint store for them.

Your first impulse may be to paint your fire engine with the finest lacquer paint you can buy. If you are going for a 100-point show truck, this is fine and lacquer is one of the easiest paints to apply due to its fast-drying characteristics. However, if your vehicle is going to be more of a working truck, remember that taking the family for rides, driving in parades and participating in fire equipment musters will take their toll on fragile or exotic paint jobs (Fig. 5-37). You should therefore probably elect to use one of the more durable modern acrylic enamels on the market today. When combined with a bit of extra hardener that is available at auto paint stores, the finish can be polished to a lacquer-like finish. You might also choose one of the even harder *Imron* or *Centari* paints that are almost impervious to chipping and scratching (Fig. 5-38). Most new fire engines are being painted with this type of paint for these very reasons.

Fig. 5-37. It is likely this 1920 Oldsmobile ladder and chemical truck is painted in a high quality modern enamel. The owner can have fun with the truck without being unduly worried about the finish.

Fig. 5-38. A modern truck like this cab-forward Mack pumper from Pomona, California, most certainly carries a chip-resistant hard paint such as Imron or Centari.

In any case, whatever paint you choose, follow the manufacturers instructions to the letter. If you have never painted before, practice on the parts of the truck that will be hidden until you master the techniques. Most of all, be sure to paint in a clean, dust-free and well-ventilated garage. You may even find a local body shop that will permit you to rent its paint booth some Saturday or Sunday when it is normally closed.

You might also want to consider having the local body shop actually apply the paint after you have done the preparation, masking, sanding, etc. A greatly reduced price can often be negotiated this way. A final alternative may be to have a body shop do all of the paint and body work, while you proceed on areas where your confidence is greater.

You should count on chipping and scratching at least a few parts during the reassembly of your fire engine. Chassis and driveline parts can be touched up with a fine brush, but if you are unfortunate enough to mar a major body part, you may be forced to spot in the panel or repaint it altogether. In any case, don't get too discouraged by these minor setbacks as they are common to even the most professional restorer (Fig. 5-39).

Fig. 5-39. A gleaming paint job such as the one on this 1920 GMC will be the crowning touch to your fire engine restoration.

UPHOLSTERY

Before painting, examine your vehicle's upholstery carefully to see how it is attached to the seat. On some trucks it is easier to do the reupholstery before painting and then to mask off the new material. On other trucks, just the reverse is true. Most restorers prefer the latter course of action so that expensive leather is not endangered by paint overspray or runs.

Once again, the upholstery (Figs. 5-40 and 5-41) of your fire engine should be no different than any other vehicle of its era. The basics are the same. On earlier fire trucks you will find beautiful button-tufted leather seats. As you get to more modern trucks, you will find simpler vinyl seat coverings finally evolving to plain vinyl-covered bench seats. The latter should not prove to be difficult for even the most rank amateur upholsterer provided you can sew in a straight line and have access to a sewing machine that will penetrate heavy vinyl or leather. The button-tufted type seat is not totally out of the realm of the amateur by any means. As unlikely as it seems, there have been some very good how-to articles and books produced by hot rod publications on this topic because that type of upholstery is very popular among the rodders.

Be sure to disassemble the seat carefully. Examine every fold and insert of the material as each will be important in determining how the seat will look and rise.

You will most likely have rebuilt any wooden parts of the seat that may have needed replacement, so now is the time to examine the springs and areas that are tied together. If the seat has been open to the weather for too long, you may want to get it sandblasted and repainted with a durable black paint. Look for areas where the seat has weakened or is binding and rebuild these areas. Your local upholstery shop can tell you where to purchase new seat springs in almost every size and shape. Once again, the springs and body of the seat are areas that will never show once the restoration has been completed. How these parts look won't be as important as how they function, although you will want the seat to look as much like the original equipment as possible. Some fire buffs who plan long rides in their trucks are not diverse to a little extra padding here and there for comfort. Again, it's your decision.

Just as with the painting, you may elect to have a professional do your upholstery, but even then you may have to help him find the correct vinyls or leathers for the job because restoring fire engine seats will more than likely be out of his normal line of work. Such leather and upholstery suppliers can be found in any issue of Old Cars newspaper or Hemmings Motor News. There are several suppliers on both coasts.

STRIPING, SCROLLWORK AND LETTERING

The latent artist shows in every fire engine collector/restorer. As you will see, the three arts of striping, scrollwork and lettering are all similar and are not difficult once you learn the basics. The keys to success are a good eye, a steady hand and patience (Fig. 5-42).

Here is where the tracings you made of the original art on the truck will come into play (Fig. 5-43). If you didn't make them before you painted the fire engine, it's too late now and you're on your own. If the detail art had already been obliter-

Fig. 5-40. The original leather interior in the American La France shown here has held up surprisingly well. It could get by with softening and dying. To replace it would require a match of the exact button-tufted pattern.

Fig. 5-41. This nicely reupholstered seat in a twenties-era GMC/Boyer has a plain back with button-tufted seat. It has been redone in a more durable vinyl.

ated or painted over before you bought the truck, perhaps any photos or blueprints you may have collected on your particular vehicle will indicate the location, size and form of the art. From this you should sketch the full-size lettering and other details on a paper or tracing paper background, just as though you had traced them from the truck in the manner mentioned earlier.

After you have a layout out on the lettering, scrollwork, etc., in actual size, the back of the paper will have to be coated with something that will transfer to the truck's finish. Sign lettering shops or art supply stores should be able to provide you with a chalk-like substance that will do the job. Simply coat the back of the paper with the chalk, carefully tape the paper to the truck in the location you want the lettering to

Fig. 5-42. True artistic talent will be needed to duplicate this intricate scroll pattern.

Fig. 5-43. This unusual lettering style on Sandy Hollow would be difficult to reproduce without a good tracing, or at least photos of the former design.

appear and then draw over each line on the paper that you wish to appear on the body. After you remove the paper, you will have a perfect image on the vehicle to work from. Of course this step should never be attempted until the finish coat of the truck's paint job has hardened completely, and any waxes, polishes or other material removed.

The first step in the process of gold leafing is to paint the lettering, scroll or stripe in a clear substance called *sizing* (Figs. 5-44 and 5-45). This colorless paint is almost the same as sign painter's enamel. This sizing should be applied just as if you were painting the details in color. In fact, many professional sign painters prefer to use white or yellow sign lettering enamel instead of sizing in case there are any breaks or open spots in the gold leafing after it is applied. This lighter base coat prevents the red paint of the truck from showing through the gold leaf. If you're unsure of which technique will work best for you, paint a spare piece of metal with the same paint as on your fire engine's body and experiment with both types of background to determine which you prefer. This

Fig. 5-44. Sizing is applied to the door just as though it was being lettered in colored paint. The powdery substance is a zinc-oxide powder used to make the sized area show up better for gold leafing and to make the sizing adhere better to the hard Imron paint.

Fig. 5-45. This close-up shows the clear sizing after it has been lettered onto the door of a new I-H/Pierce pumper.

sample metal can also be used to experiment with the actual painting of the lettering and other details, as well as color combinations.

When the sizing or background enamel has set up to the point of stickiness you are ready to proceed with the application of the gold leaf.

Gold leaf should be available from most sign painters' supply houses or similar businesses serving the commercial art field. Made from real gold, you will find gold leaf amazingly fragile and expensive. It is normally available in sheets or on rolls in a tape-like form. For small areas, working with the sheets seems to be best for most people. The leaf is picked up with a brush and applied to the areas to be covered (Fig. 5-46). At no time should your fingers touch the leaf because they will leave fingerprints. Remember that the leafing is fragile and anything hard will mark it. This fragility is precisely why gold leafing is used to highlight details on a fire engine's art, but if used carelessly, it will work against the restorer. After the leaf has been brushed on, lay a piece of clean paper over it and press it gently into the sizing. If using the gold leaf tape roller method, the roller will perform both tasks of application and fixing the gold to the sized areas. Gently brush away any leaf not attached to the truck via the sizing.

You might desire an engine-turned (Fig. 5-47) or swirled effect in your gold leafing. Once it has set up a bit with the drying size or background paint, take a cotton ball between two fingers and in systematic progression, press it into the gold leaf and give the cotton ball just one twist for each swirl. It is advisable to practice this technique on your sample metal first. It is also worthwhile to note that while many modern fire engines feature this eye-catching swirled effect in the gold leafing, most older rigs carried only a flat-leafed effect.

After the leaf and its undercoat has completely set up, you can proceed to protect your art work with a coating of clear varnish or enamel, whichever is most compatible with the overall finish of your fire engine.

After this top coat has dried you will want to outline your gold leaf lettering, possibly using a contrasting color as an outline. This requires a good quality paint brush and special

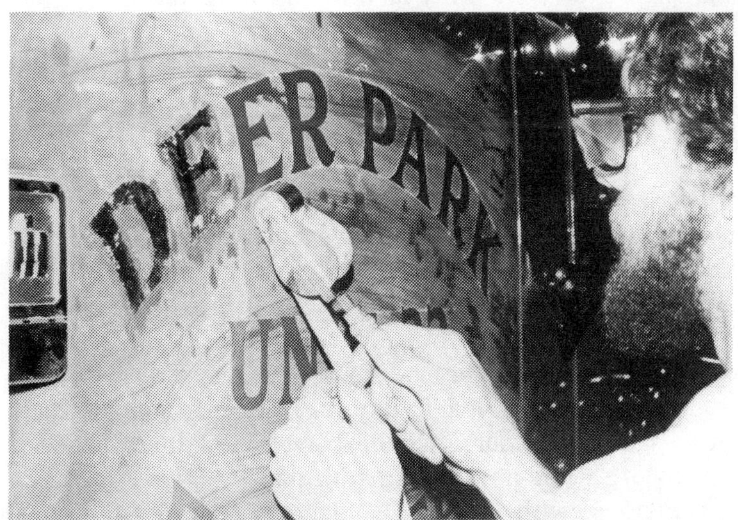

Fig. 5-46. Here we see a lettering artist hand rolling gold leaf onto the same door. Notice the flaky lifted areas in the gold leafing already done. This can be lightly brushed away later, but be careful not to leave fingerprints or to mar the gold leaf.

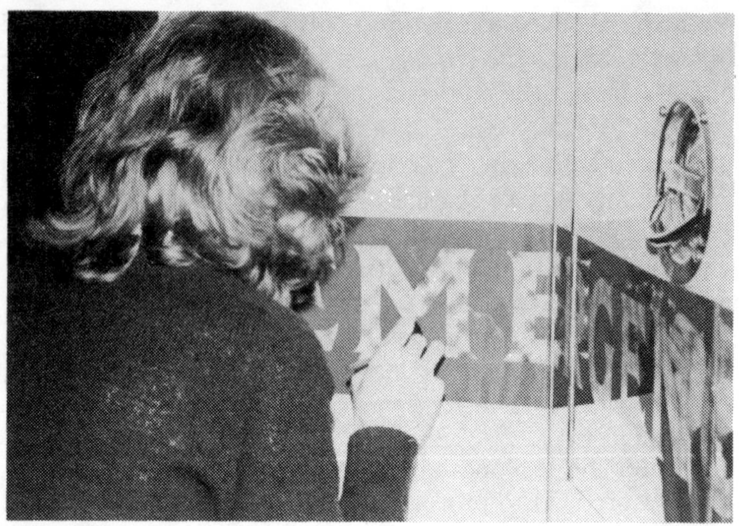

Fig. 5-47. Here, another artist applies a protective clear coat over the gold leaf on a new emergency truck in the Pierce plant. Note that this gold leaf has been engine turned.

sign painter's enamel. My favorite is *One Shot* sign enamel. You may also elect to go with a bolder shadow behind the lettering, maybe even using a third color. This requires the same type of paint and a slightly larger brush (Figs. 5-48 and 5-49).

Some of the most common combinations are gold leaf with a black outline and shadow and gold leaf with a black outline and white shadow. I have even seen a fire engine running gold leaf with a black outline and green shadow that is surprisingly attractive. Other varieties (Figs. 5-50 through 5-53) include a yellow fire engine with gold leaf lettering, outlined in black with orange shadowing and a cream-colored truck featuring gold-leafed letters, black edging and a dark blue shadow. Of course the best bet is to know what your truck wore when new and go with that.

Striping with gold leaf is done in basically the same manner as lettering, but there is an alternative. An artificial gold leaf stripe is being made by 3M for fire apparatus and can be ordered through your local auto parts store. It is extremely

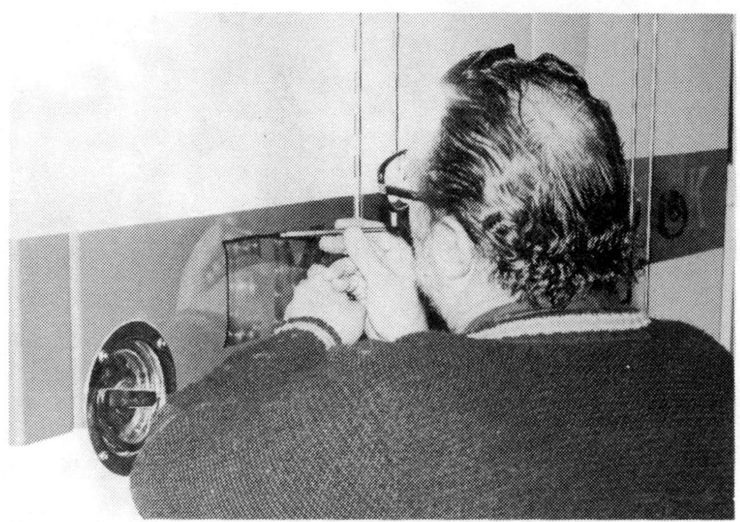

Fig. 5-48. After the clear overcoat has dried, an outline color and even a shadow is applied around the edges of the gold leafed letters. This step will help cover small irregularities in the edges of the gold leaf.

Fig. 5-49. Only the top word on this Morrison Fire Department emergency truck has been shadowed at this point. Notice how the shadow makes the word stand out more from the background color.

Fig. 5-50. This truck will carry a double shadow effect. Notice the top word has already been completed with a black shadow over a dark red shadow on the bright red truck door.

easy to apply, comes with several optional stripes and fancy end designs and is much cheaper than real gold leaf. The only problem is that on close inspection, it also looks cheaper. However, on a fun truck or parade truck, this might be perfectly suitable. There are other aids for the fire engine collector who is not particularly adept in artistic endeavors. These include scrollwork and lettering decals, although they may be a bit difficult to find. Check the fire equipment manufacturers listed in the back of this book, or consult other fire buffs for possible leads.

Just as with gold leaf lettering, pinstriping (Figs. 5-54 and 5-55) of your fire engine can be done in several different ways. One is the old sword brush technique which takes a lot of practice but is the way striping was done in the old days. You will get the most pleasing stripe this way if you have the patience to learn to control the brush wheel. Sword striping brushes are available through art supply stores and the best paint, again, is the sign painters enamel.

Beugler Stripers, P.O. Box 29068, Los Angeles, CA 90029 have been around for many years building a quality wheel-type striper. This tool was used by many fire truck manufacturers in the height of the pinstriping days of the 1920s and 1930s. For many years thereafter it was also a common tool in auto body shops. It is easy to use and the best substitute there is for hand striping. This type of striping tool runs between $30 and $65 and is well worth the money. There are a number of cheaper tools on the market, however. Even Beugler cheapened its tools to the point of being virtually

Fig. 5-51. This lettering on a truck for Fort Dodge was especially attractive, though not traditional. It is gold leafed with an orange shadow and black outline on a yellow truck.

Fig. 5-52. This Ford/Pierce was a very unusual olive green with gold leaf striping and scrollwork. Most green fire engines have a brighter, high-visibility lime green. Whatever happened to tradition?

unusable for several years. Don't compromise on a cheap striping tool because you will pay for it with a poor job.

Striping can be done with tape, as well. Some of today's tape are very flexible and can be worked around corners, etc. While not as satisfactory as the first two options, you can certainly give your fire engine a nice effect with commercially available tape. Don't buy just any cheap tape available at the local discount store. Get the better quality material offered by auto body shop suppliers.

The last form of striping is done with a special masking tape. The tape is applied to the body of the vehicle, then a small line of tape is removed from the middle of the strip.

However very few of these jobs seem to turn out nice. It is difficult to get the masking to produce a good sharp edge, and the stripe always seems to end up looking fuzzy.

Decals can also be an acceptable alternative for the non-artistic fire engine restorer. When using this method be sure to follow the manufacturer's directions very carefully and use plenty of water when applying one of the larger decals. A rubber squeegee is handy for gently working out air bubbles.

For all of these processes, you will want to work in a warm garage or outdoors on a nice day. Paint and varnishes tend to thicken, become sticky and dry much too slowly in a cold environment. Tapes get brittle and decals break apart or develop cracks. It is best to have a warm garage available to you for this type of work, and if you're bringing the fire truck in from outside, make sure it has had time to warm up to room temperature before you start.

TIRES

Tires for fire apparatus fall into several categories. The first, and most commonly associated with antique fire en-

Fig. 5-53. A close-up of the same scrollwork shows black outlining, engine turning of the gold leaf and darkened shadows on the scrollwork done in a very subtle root beer reddish brown. A true tribute to the artist and the manufacturer who still take time to letter their trucks in the old manner to insure a quality look.

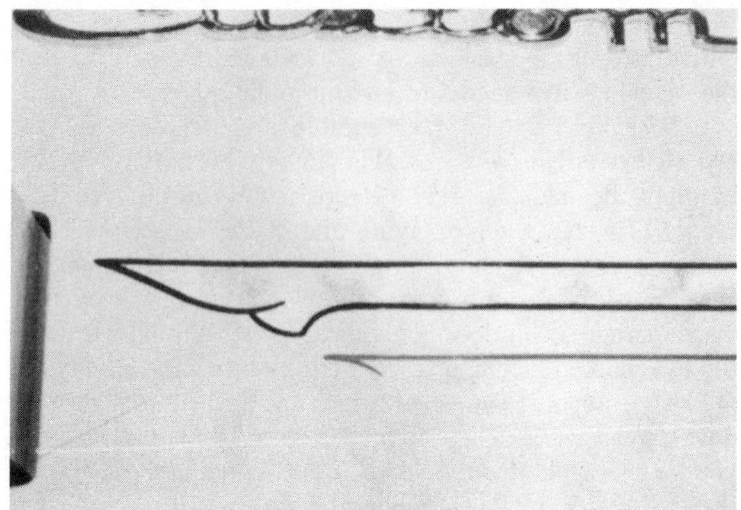

Fig. 5-54. Notice the hand-applied gold leaf stripe and hand-painted pin stripe below.

gines, are the *hard rubber* tires composed of solid rubber (Fig. 5-56). They are similar to what you may have seen on a fork lift truck in a factory, or even on a child's tricycle. In antique

Fig. 5-55. Now compare commercial #M automotive tape gold striping and end, plus white tape pin stripe below. Quite acceptable, but just not the same as the real thing.

Fig. 5-56. This early Jeffrey Quad from the Whitewater, Wisconsin, Fire Department is a classic example of a fire truck riding on hard rubber tires.

hard rubber tires, the most commonly found sizes on fire equipment are the 22 inch and 24 inch varieties.

Solid rubber trucks are found on almost no fire trucks today. They were universally replaced by fire departments as soon as confidence was developed for the pneumatic balloon tire. Those trucks which had originally been equipped with wooden spoked wheels usually had them cut down and fitted with 24-inch rims common to the trucks of today. This poses two problems for the restorer: rebuilding the cutdown wheels and fitting the new old-style rims with hard rubber tires.

After a long search, you might find a set of original, correct wooden wheels, but they will most likely be dry rotted and useless. Don't trust the safety of your truck to marginal wooden wheels. They take all of the punishment and the front wheels are prone to collapse the first time you hit a chuck hole or a corner or sometime when most of the g-forces are pushing against a sideways turning wheel. This is not an area for guesswork.

If you are a competent woodworker, go ahead and re-wood your wheels. If not, enlist one of the few men who are. Wheelwrights today are few and far between, but several advertise in Hemmings Motor News and Old Cars Newspaper.

People who will fit your wheels with solid rubber tires are even scarcer, but they also advertise in the collector car publications and fire buff magazines. It has only been very recently that collector demand for solid rubber tires has reached the point where they can be relatively easily obtained. Restorers of just a few years ago had to beg rubber companies to do a special job for them, or get an industrial truck tire dealer to try it. Some even resorted to applying bands of rubber to the rim and vulcanizing the seam.

Another interesting alternative may be open to you if you live in any part of the country which has a near-by Amish community. Because these religious people will not own modern means of transportation, virtually every settlement of Amish has a talented, honest wheelwright less than a day's buggy ride away. By now, most Amish craftsmen are used to requests from outsiders to do special jobs. While they are not interested in such frivolitries as antique fire engines, they won't begrudge you of yours and they will usually be able to produce a set of wheels to your specifications. Even though his shop looks like it might have been there in the 1890s, don't expect to get your wheels built for a century-old price. Though the shop may be heated with wood and lit by kerosene, don't be surprised if the wheelwright figures out the estimate for your job on a battery operated calculator.

Antique tires are very hard to come by (Figs. 5-57 and 5-58). They are currently being offered by only two United States tire dealers. They both advertise in the fire buff magazines and collector car periodicals. You will find these tires very expensive so when you go to look for a fire truck, weigh very carefully the condition of its tires when figuring what the vehicle is worth. It is not uncommon for one new tire to cost

Fig. 5-57. While not a motor driven piece of apparatus, this early steam pumper has wooden wheels of about the same construction as the motorized units and would pose the same restoration problems.

Fig. 5-58. Pneumatic tires on a small Ford truck like this one would only run a 30 × 3½-inch tire on the front, a widely produced size available from several antique auto tire dealers. Large tires of the 23-inch size are extremely hard to come by and tire dealers take advantage of the situation with exhorbitant prices when they do make them available.

over $200 and these are stories of a set of tires with matching tread pushing the $300 mark for each. Also be aware that the tires come in several degrees of quality.

Of course the best you could hope to find would be a new set, but there are no new tires in these sizes being manufactured in the United States today. There are rumors that such

Fig. 5-59. The massive bumper, headlamps and radiator shell on this Pierce-Arrow fire engine will prove very costly for the owner to restore unless he owns his own plating shop.

Fig. 5-60. An item like this siren will have to be completely disassembled to be re-plated. Each screw and part will have to be inventoried, or better yet, laid out on a clean sheet and photographed before going to the plater.

tires were being built up until just a few years ago in England, and it is not uncommon for storehouses of them to still exist. Likewise, the retread or remolded tire is likely to be an import. Highway tractor trailer trucks almost all use retreads and if well done, they are a good alternative to a set of new tires. It is also a good bet that you will get a set with matching tread design this way. Just be sure to examine the bond between the tire and tread well and look at the overall condition of the tire carcass. Another cheap solution to the tire problem is used tires. Several of the tire dealers who handle truck tires might be able to help you in this department. Once again, examine the tires carefully for cuts, weather checking and general deterioration. If you don't care at all if your tires match, this will be your most economically sound approach. The most prevalent size tires (for 24-inch rims)are the 36×6, 38×7 and 40×8.

Modern size tires in the 20 inch range are offered by almost every major new tire dealer in your area. You can consider yourself fortunate if your fire truck is equipped to run on tires of this size.

If you have purchased a fire truck that was retrofitted with pneumatic tires in place of original hard rubber equipment at some time in its career, consider leaving them on. They offer the advantages of a softer ride and better handling, not to mention the differences in cost in having wheels rebuilt and purchasing solid rubber tires. Unless you are building a 100-point show truck, you will most likely be happier riding on air.

PLATING

Replating the metal parts of your fire engine will probably be one of your most difficult challenges during the

Fig. 5-61. Many older trucks may look out of place with high luster chrome when original specifications called for the duller nickel plating. However, for musters and just fun, the chrome is a more durable and maintenance-free finish.

Fig. 5-62. Many fire trucks carried a large amount of nickel plating. Many bells like the one on this American La France were nickeled.

restoration (Figs. 5-59 through 5-66). It is the one job you will most likely be unable to perform yourself. Also, good platers are harder to find than any other service your fire engine is likely to need. But if you live in a larger city, it is possible there will be several plating shops right in town. Upon visiting them, you will find them to be production shops. That is, they are interested in plating 100,000 kitchen chair legs for a local furniture factory and could care less about doing the one nickel plating job for your fire truck's bell. Check with any local antique car restorers you may know to find out if they have any sources for their plating work. There are several platers advertising in collector car publications, but mailing the bumper of a fire engine can be an expensive proposition and there is always the chance of loss in transit of an irreplaceable part. If at all possible, even if you have to deliver the parts over a long distance, take them yourself. The personal contact with the plater will be worth the trip in terms of the suitability of the finished job.

Before you leave the garage, inventory the parts you are taking to the plater. As mentioned earlier, it is also a good idea to photograph them. It is very common for parts, especially small parts, to drop to the bottom of the plating tanks. And as stated by a popular variation to Murphy's Law, if the plater is going to lose one your fire engine's parts, it will the most valuable or least easily replaced part.

Fig. 5-63. Some radiator shells are actually the outside covering of the radiator and cause great plating problems because the radiator has to be totally disassembled to be plated.

Fig. 5-64. A heavy aluminum casting like the hood ornament on this pumper can be polished again by a plater or on your own buffing wheel to a chrome-like finish.

Ask the plater about the levels of chrome and other plating he offers. Assess these levels with an eye toward your budget and the degree of restoration on which you have settled.

Many old fire trucks used nickel for most of the plated finishes. For that reason, the most ardent perfectionist will want nickel plating on his truck. However, if you are going to be using your truck often, be aware that nickel tarnishes easily, picks up fingerprints like magic and costs more than chrome. Chrome, while it may not be strictly authentic for your truck, is a flashier trim, provides greater durability and ease of care and is less expensive. So, as in virtually all matters of the restoration of your fire engine, the choice is yours.

Be sure to pin your plater down to a reasonable delivery date. Ask around about his reputation for making promised deliveries and make him stick to his promises. This might require several calls during the waiting period, and possibly even a personal visit or two. Make him aware that you expect him to honor his deadline and that you will hold him responsible for any missing parts. This sounds harsh, and

Fig. 5-65. This brass Pierce-Arrow hood side panel emblem was also polished rather than plated.

Fig. 5-66. The pump unit on this American La France pumper has dozens of small parts to be plated. Careful disassembly and inventory will assure safe return of all your parts.

though you might not like playing the heavy, this attitude is a must. It is also the reason the production shop platers probably won't want anything to do with you. Therefore, you will probably have to search out one of the few specialists serving the hobbyist.

You might be able to work out an arrangement with the plater to do some of the work yourself, especially the hand work. For instance, you might have him chemically strip the old plating from your parts while you do the hand buffing to prepare them for the new finish. Don't attempt this without his full cooperation and without knowing precisely what you have to do to the part before you return it to him for the new plating. Otherwise you are unlikely to get the finish you are looking for. Second only to the paint and striping of your fire engine, the quality of the plating will determine how well your truck looks overall. Don't cut corners here.

6

Exhibiting Your Antique Fire Engine

In recent years commercial vehicles have been infiltrating the antique collector car circles in ever greater numbers. Relatively easy to restore in comparison to the more delicate automobiles of the same era, antique trucks have opened up a whole new field of motor vehicle sport for collectors. Naturally, fire engines play a very important role in the commercial vehicle classes at auto shows.

ANTIQUE AUTO SHOWS

The collector who decides to purchase and restore his truck for show purposes only will have different motives and goals in his restoration (Figs. 6-1 and 6-2). His gold leaf striping may be flawless, but his pump may be non-functional. It is interesting to note that at fire equipment musters the emphasis is exactly the opposite. More interest is in how the truck can perform the jobs it was built for rather than how pretty it is.

Car show judging, unfortunately, is not always the impartial process it should be. At its worst, it can range from incredible nitpicking by a team of nevertheless qualified experts to a pretty unfair process run by self-appointed experts who are swayed by personal likes and dislikes in motor vehicles and even friendships with the owners. At any judged

Fig. 6-1. This highly restored 1916 Model T Ford owned by Alan C. Myers has been admired by thousands of spectators at many eastern car shows.

show you will always have a few happy people and a larger number of unhappy participants. Judging sometimes creates this type of atmosphere. It is unfortunate for the hobby, but understandable when you consider that such feelings will arise any time someone puts years of work and perhaps thousands of dollars into an effort to win a $5 trophy and the approval of a select group of judges.

Most car shows are judged on a 100-point system with 100 points being the best possible score. Vehicles seldom reach this level. Judging is based on the quality of restoration and thus vehicles are sometimes over restored; that is, restored past the point of originality. Most shows require that the vehicle is drivable and that it be operated at the show. In the name of preserving a fine restoration, it is not uncommon for the vehicle to be trailered to the show site and driven only a few feet past the judges.

One of the largest and most respected judged antique motor vehicle shows is held at Hershey, Pennsylvania, each fall in conjunction with the largest swap meet in the world. It is sponsored by the Antique Auto Club of America. Other significant judged shows include the Pebble Beach, Califor-

Fig. 6-2. Another fire truck at an old car show, this 1937 Hale owned by Howard Daniel is certainly imposing-looking around the smaller cars.

nia, Concours; the Ambassador Concours in Los Angeles; the Hoosier Auto Show in Indianapolis which includes a fire muster every year; Stone Mountain, Georgia; and many others.

Fig. 6-3. The non-judged type car shows like the one held in Iola, Wisconsin, each July are becoming more and more popular due to the relaxed atmosphere. Fire trucks are always welcome.

Fig. 6-4. Here we see firemen from several Wisconsin small towns battling a beer barrel at the annual Bear Creek Sauerkraut Festival. Muster games have been part of the fun there for years.

Fig. 6-5. Same game as in Fig. 6-4 with different fire fighters at the giant muster at Greenfield Village, near Dearborn, Michigan.

Fig. 6-6. Firemen scramble to hook up the hoses to a vintage Seagrave pumper in a contest to knock a basketball from the window of a fake house not shown in the photo. This action was at the Indianapolis fall muster recently.

Another type of car show, and one that is becoming increasingly more popular is patterned after the giant show held in Iola, Wisconsin, each July by Old Cars Weekly newspaper and various civic groups in the town (Fig. 6-3). The concept is that there is no judging. It is strictly a fun day in which participants are admitted free for driving a collector vehicle to the show. A companion swap meet and chicken roast help draw crowds and raise funds for the civic groups' work. The concept has proven to be so much fun for participants that the Iola show has grown to nearly 2,000 collector vehicles each year. Other shows using the same format are becoming more and more popular.

FIRE ENGINE MUSTERS

Emphasis is on performance, not showmanship at fire engine musters. However, this is not to say that some incredibly well-restored trucks are not present or that the men do not take as much pride in their vehicles' appearance as the antique car restorer.

Fig. 6-7. For the next step in the same game as in Fig. 6-6 firemen shoot water through the open windows of the fake house as though it were on fire.

Fig. 6-8. Taking in water from a nearby stream and pumping it back out again at the Greenfield Village muster.

Fig. 6-9. Hand cranking up the old ladder at the Greenfield Village muster as visiting firemen look on. It isn't often that you see the sight of several hook-and-ladder trucks, all with their ladders extended in the same spot at a muster.

In fact, at some musters the trucks are judged on appearance and authenticity just as at a car show. However, the trucks are also put into competition. It is not uncommon for active fire departments to participate right along with the hobbyists (Figs. 6-4 through 6-9). Engines will include the very latest in new equipment as well as the antiques.

Fire muster games include *First Water*. This game is just

Fig. 6-10. Exhausted fire fighters relax near and against an early post-war Ford pumper and enjoy the break at the Indianapolis fall muster.

what it says—a race to see which pumper can produce a full stream of water in the shortest time. Another game is a sort of tug of war with competing pumper crews using their hoses to try to force a beer keg hanging from a high wire over to the opponent's support pole. In other games, fake fires are extinguished where trucks race to a hydrant to hook up and pump water, then knock a basket out of the window ledge of a makeshift building.

Some of the largest musters in the country are set in southern California, Syracuse, New York and Dearborn, Michigan. Other musters are held across the nation and many host communities sponsor one or two events of this nature in conjunction with a local fund-raising event.

In any case, fire equipment musters provide action unparalleled in the old car hobby (Fig. 6-10). Manpower and teamwork considerations are placed above those of personal or vehicular vanity.

PARADES AND TOURS

No parade would be complete without a string of fire

Fig. 6-11. This terrific fire apparatus parade is at the Greenfield Village muster. Local parades with just a few trucks and vintage cars can be almost as exciting for the enthusiast and definitely as exciting for the crowds.

Fig. 6-12. Ready for a parade, this ladder truck has been converted to carry a long bench inside for passengers. A fold-down step has also been added for ease of entry for the riders.

engines roaring down the street with lights flashing and siren and bell sounding. This type of event is possibly the best opportunity for you to share the pride and pleasure of antique fire truck ownership. Children and adults will want to ride with you. Civic groups may call on you to haul dignitaries or parade queens. Besides, parades are just plain fun and American as apple pie. It's a good idea to make sure your truck is insured for such use and that its cooling system is in good order because the slow pace of a parade can be hard on many older vehicles. Fire trucks generally seem to hold up well under these conditions, however, since their cooling systems were built quite large to stand up to hours of pumping at a fire. So don't hesitate to participate and enjoy yourself (Figs. 6-11 and 6-12).

Tours are more common to antique car clubs, but are not unheard of for vintage trucks and even especially for fire apparatus. In almost every case, you and your vehicle would usually be welcome with any type of tour group. Usually

Fig. 6-13. Just the type of truck you are likely to see in a fire museum rather than on the street is this 1889 American La France steam pumper mounted on a 1916 Seagrave motor tractor.

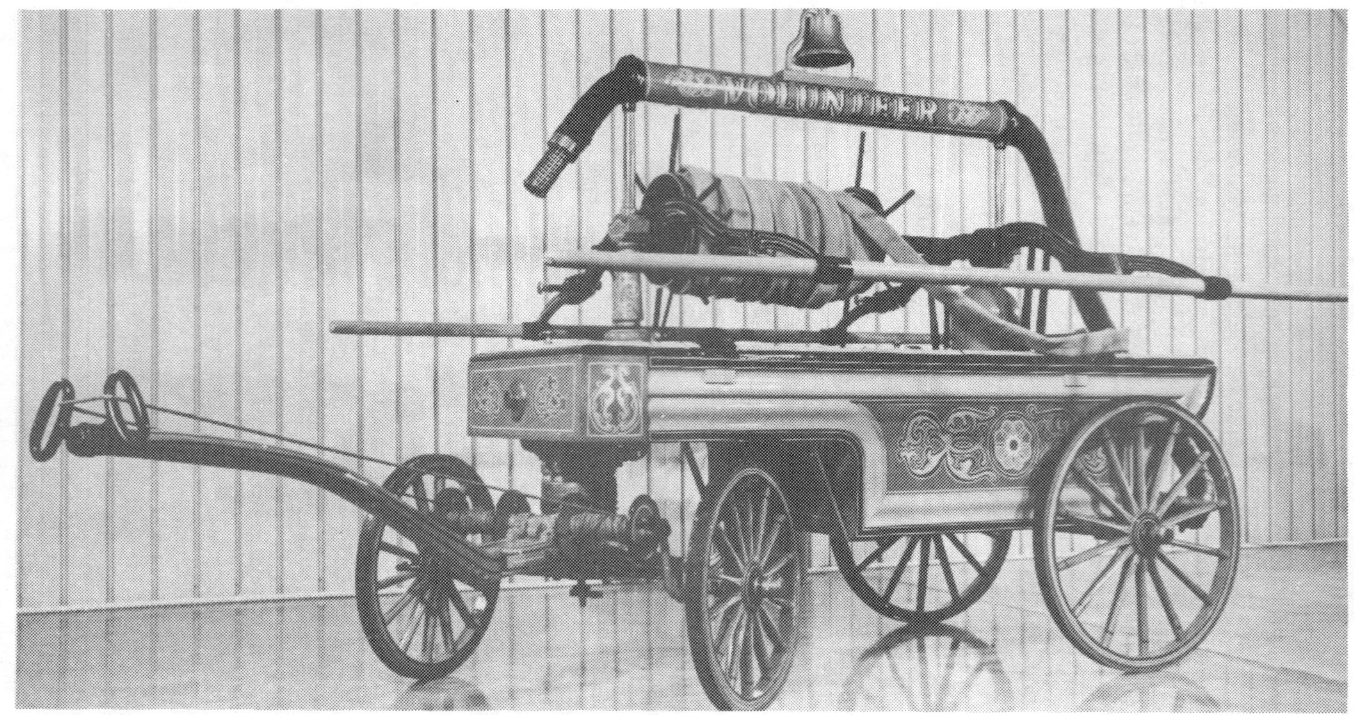

Fig. 6-14. Part of the exhibit at the Hall of Fame Museum in Phoenix, Arizona is this Rumsey Hand Drawn Pumper, Volunteer.

planned around a Sunday afternoon, the tour will consist of a picnic lunch and a ride in the surrounding area, perhaps to some local point of interest. If you are lucky, maybe the tour will visit a nearby firehouse, museum or even include a visit to a fire equipment factory. These events are low-key and a lot of fun as well. Of course, anyplace you find old car enthusiasts and their families, you will likely be obliged to provide a few rides in the fire engine during the tour. Somehow the magic of a fire engine ride is special to enthusiasts of all ages, and of course it's pretty hard to say you don't have room.

FIRE MUSEUMS

There are many more fire museums in this country than can be listed here but some can be found in the appendices. Many are run by private buffs and collectors. Some are actually private collections called museums for tax or licensing purposes, but most are always open to the true enthusiast with a phone call ahead. The major museums usually have a full complement of memorabilia and engines from early steam pumpers to more recent giants (Figs. 6-13 and 6-14). This is a good opportunity for you to see trucks in a restored state and judge what you have to do to your own. You will also see a progression in the manner in which fires have been fought through the years in this country. You will be amazed to see tiny hand pumpers, buckets and wooden ladders that were used many years ago. If the museum is run by a local fire department it is most often near the town's active fire station and you can compare the vintage equipment with the modern trucks housed nearby. While your wife usually hopes your family vacation might be free from hobby interests, this is a great time to take in a museum or two out of your home area.

7

Fire Memorabilia

There hasn't been a Christmas in the last 100 years that some child has not received a big red toy fire engine from Santa Claus. Maybe if he was lucky, the fire engine was big enough to ride in and peddle all around the neighborhood. These, of course, are the memories that make a fire buff, especially when he remembers sitting in his peddle truck on the sidewalk as one of the real machines roared and screamed by on a life-saving mission.

TOYS AND MODELS

These same toys of childhood fantasy are today collected by fire buffs and toy collectors alike. Including cast iron toys (Fig. 7-1) made in the image of horsedrawn steam pumpers and ladder trucks, these toys bring bigger and bigger prices each year. Some of these early toys are even reproduced so well that it is hard to distinguish the copy from the original. When purchasing such toys from an antique dealer or a flea market vendor, be sure you are not getting a modern replica that has been artificially aged to look authentic. If you are merely seeking decoration in a fire motif these reproductions are not only adequate, but may be preferred due to their lower cost.

Tin toys such as Buddy L, Tonka, Kingsbury and the like

Fig. 7-1. Toy fire apparatus goes back to before motor-driven trucks were used, as these old cast iron toys indicate.

Fig. 7-2. An excellent comparison between the early fifties cast aluminum Smith-Miller Mack hook and ladder and a less desirable Tonka steel hook and ladder from the late fifties.

Fig. 7-3. In the foreground is a fifties-era Tootsie toy emulating a closed cab Mack. To the rear is an AMT American La France Ladder Chief assembled in 1/25th scale. It is not a collectible yet, but fun to build. The small truck is a Hot Wheels Chevrolet rescue truck.

are possibly the most in demand. These are the toys best remembered by buffs from their childhood days. Even companies like Marx, which produced a lower quality toy than those already mentioned, are still collected with a passion. In the same vein as the tin and metal trucks, Smith-Miller and Smitty trucks are the most highly-prized and usually the most highly priced. These trucks were super high quality cast aluminum toys actually used as promotional models by some Mack truck dealers. Produced in the late 1940s and early 1950s, it is not uncommon to see a prime Smith-Miller Mack fire engine toy sell now for over $200 (Fig. 7-2).

With the advent of plastic, lower priced toys became the wave of the future in the 1950s. Marx, Gay and other manufacturers produced dozens of fire engines for boys and girls. The American LaFrance 700 series cab-forward truck has been reproduced by so many manufacturers that it has become the standard toy fire truck design.

Recently, I took a novice to a fire muster. His comment was that he really liked the ALF 700s because they looked like big toys. We are not sure that ALF would appreciate the compliment, but it really is a compliment to the classic design that these trucks were selected by so many toy manufacturers right down to this day.

A trip to the local dime stores still reveals toy fire engines in every material and price, just waiting for the child's touch. These are the collectors items of tomorrow. Many fire buffs have large toy collections and attempt to keep them current by purchasing new toy fire trucks as they come out.

Corgi, Matchbox and Dinky are just a few manufacturers who specialize in the popular small (1/43-scale) die cast toys. These toys closely approach being models and are relatively expensive when compared to their plastic and tin counterparts. On the market since the early 1940s, these toys have become among the most collectible. The most common die cast toys in the United States are Mattel's Hot Wheels series. They have not slighted fire buffs, either, with several fire-oriented toys in the line (Fig. 7-3).

Oddly enough, model fire trucks have proven to be less popular than the toys and fewer have been produced. Lindberg and Aurora produced some of the earliest kit fire engines, once again choosing ALF trucks. In the early 1970s, AMT introduced an excellent line of 1/25th fire apparatus including new ALF pumper, snorkel and ladder trucks, a Chevrolet ambulance and a Chevrolet fire chief's car. The line was discontinued over the years so that today only a cheapened version of the Ladder Chief aerial ladder truck is available. Original older kits, still in unbuilt form, already command high prices from collectors. Gabriel currently produces an excellent metal 1909 Ford Model T chemical truck kit. Model builders with a little prior experience should also look into the super small scale kits, such as those built for HO-scale train layouts. There are several good 1/87th fire engine models such as Ahrens Fox, Mack and Ford and there are even scale old-style firehouses to build with the trucks. Model building can provide many hours of fire engine enjoyment in the winter days when working out in a cold garage on a cold life size fire engine doesn't seem so appealing (Fig. 7-4).

Fire buffs are often so dedicated that they will collect

Fig. 7-4. Model building is not limited to the kits alone. Scratch-built models like this 1939 Ahrens-Fox pumper show what can be done with a little skill and a lot of desire. The model was created by Tom Showers.

Fig. 7-5. Another scratch-built model by Tom Showers shows a very rare truck—a 1936 Ahrens-Fox sedan cab pumper owned by Pawtucket, Rhode Island.

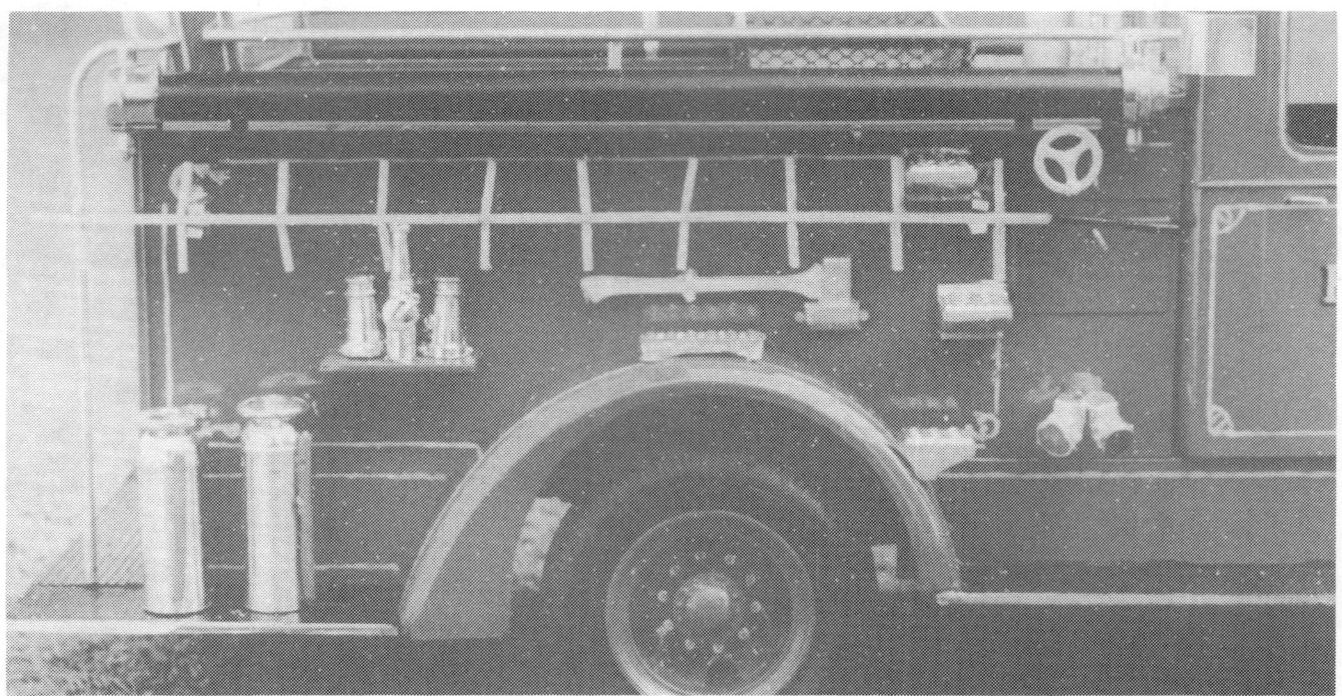

Fig. 7-6. Here is modeling at its best. This truck is in the very small 1/32 scale, or about 2/3rds the scale of the current 1/25th kit trucks.

Fig. 7-7. A close-up of the detail obtained by Tom Showers on the Fox models. Note the name plate directly under the pump unit.

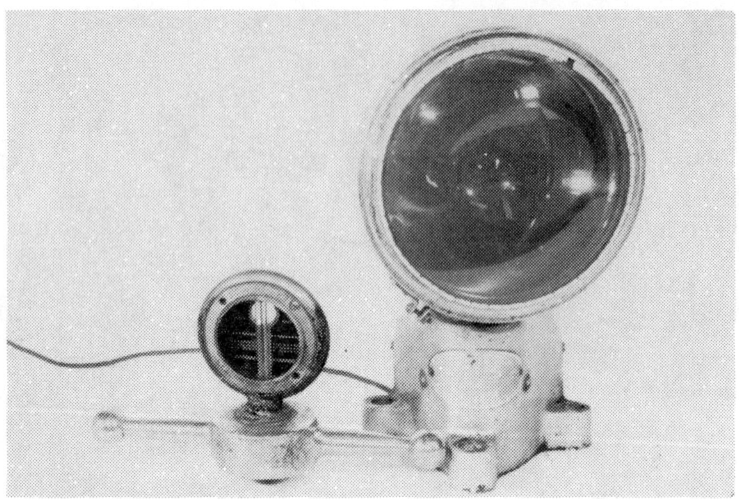

Fig. 7-8. Memorabilia can include truck equipment like the Mars light on the right and the Seagrave Moto-Meter on the left. Restored, these treasures make fine shelf pieces.

Fig. 7-9. Many collectors collect radiator emblems or medallions. This Stutz piece would be a prize for any badge collector.

Fig. 7-10. This copper oiler, polished to the hilt, will make a spectacular shelf item for the collector.

Fig. 7-11. This drum search lamp and bell are also typical of the fire engine accessories that are popular memorabilia items.

Fig. 7-12. Non-truck collectibles most certainly include the hand pumper on the left and the complementing hose cart at the right.

Fig. 7-13. This hand-drawn chemical tank unit would also make a nice collectible when restored.

virtually anything that has to do with fires, firemen and fire-fighting equipment (Figs. 7-5 through 7-17). Starting with the trucks themselves, bells, sirens, Deitz Fire King lanterns and brass hose nozzles top the list of collectibles. Hardly a fire buff stops purchasing such items at swap meets until the day comes that his truck is fully equipped. Brass and copper fire extinguishers are made into spectacular lamps that can be polished to the hilt. Fire station equipment such as call boards and brass poles are installed in family rooms. Call boxes from the city streets, and the relays they are tied into, are among the most highly desired collectibles of this type. Axes, lanterns, call box stations and just about everything associated with fire fighting in the past is actively bought and saved by buffs today.

Fig. 7-14. Literature collecting is big in the antique car hobby and the same holds true for fire historians. Original sales literature is collectible from the original old books like the 1930 ALF piece on the right, to the current Pierce book in the rear.

Fig. 7-15. In the same vein as literature collecting, original factory photos like this one showing a new 1923 Reo/Waterous just before delivery to Greenleaf, Wisconsin, are popular among fire buffs.

Fig. 7-16. Recommended reading for anyone with an abnormal interest in restoring, owning or just knowing about fire engines.

Fig. 7-17. Bubble gum cards have become very popular collectibles in many memorabilia markets and firebuffs find them just as fascinating.

HISTORIANS

Along with this passion for memorabilia, the buff becomes an historian. As he researches his truck and its home community, he finds about about the great fires of the era and area. He may collect photos of the fire stations, get plans of the fire department's call box system, learn of great fire fighters who served the community and maybe gave their lives to save others. Sometimes the interest bridges over to such areas as fire boats, airplanes, forest fire-fighting equipment and even military vehicles and history.

It is clear that once hooked the fire buff usually jumps into the hobby with both feet. His interest grows until he learns every phase of fire fighting, fire fighters and their equipment. By enjoying his equipment to the fullest during and after its restoration, he shares in the illustrious heritage of the machines and the sometimes glorious days gone by of the men who served with them.

Glossary

aerial ladder truck: Truck used for high elevation ladders, usually mounted on a rotating platform. Ladder can be from 50 to over 250 feet long. The aerial ladder truck can be a straight chassis or an articulated chassis and is then referred to as a hook-and-ladder truck.

ambulance: An emergency vehicle used to carry the injured or sick to the hospital. This unit can be based on a large luxury car chassis, light truck or van chassis, or in the case of the larger rescue trucks, heavy duty commercial truck chassis. Modern ambulances carry almost every life-saving device found in a good clinic.

artillery wheels: Wheels in which the hub is attached to the rim via heavy spokes made of hard wood or cast metal.

bell: Large metal emergency signal attached to most old fire engines. Rang manually or by electric clapper. Now making a comeback as ornament on modern-made fire engines.

booster tank: An auxiliary tank from 50 to 500 gallons on a pumper truck to supplement water drawn from a hydrant or tanker.

brush fighter: A small truck, usually a combination, with the versatility for fighting fires in rough terrain. These light duty trucks are often equipped with four-wheel drive for off-road work.

cab forward: A truck in which the cab rides forward of the engine. A design pioneered in the late 1930s by American La France.

canopy cab: A semi-closed cab truck in which the rear of the cab facing the truck body is open. Frequently has several rear-facing seats.

call box: A remote unit located on a street corner for calling in a fire alarm. Some units have a telephone inside, some simply trigger an alarm at the fire station.

chain drive: A truck in which the rear axle is driven by a chain, rather than the modern drive shaft. The chain is usually driven through a differential a few feet ahead of the actual driving axle. This differential is usually driven via a shaft to the transmission.

chemical truck: A truck carrying chemical fire extinguishers to fires on which water would have no effect, such as other chemicals, fuels, etc.

chief's car: A passenger car utilized by the chief for daily and emergency travel, and for directing operations at the scene of larger fires. Usually equipped with red lights, siren, etc.

combination: A truck that is capable of doing service in several combinations; i.e., a pumper/hose truck, a pumper/hose/ladder truck, etc.

commercial chassis: A fire engine based on a commercial brand chassis such as Ford, Chevrolet, etc., that does not manufacture fire equipment itself.

cowl-mount pump: A pumper in which the pump is mounted behind the firewall between the engine of the truck and the driver. Not a successful design, but used on many American La France Series 400 trucks.

crash truck: A fire truck used primarily at an airport for air disasters. It is usually a very large truck with four-wheel drive and large chemical-carrying capacity for fighting high octane fuel fires.

custom chassis: A fire engine that is manufactured solely by a fire engine manufacturer, not utilizing any commercial chassis as a basis. These trucks may, however, carry components built by other manufacturers. Examples include American La France, Seagrave, etc.

deluge gun: (water cannon) A high pressure nozzle usually mounted directly on the truck or on an attached boom or tower for unmanned close-in fire fighting.

disc wheels: Wheels on which the rim is bolted to the brake drum/hub. The most common type of wheel on modern trucks.

double ignition: A truck engine that has two sets of ignition components, spark plugs, etc.

emergency truck: The largest type of ambulance; virtually a hospital on wheels. Some such vehicles do not have the capacity for carrying patients and require an ambulance as back-up.

extinguisher: A small hand-held unit for battling small fires. Extinguishers come in all types of chemical combinations for different types of fires. Most often used are soda/acid, soda/water and foam.

fire boat: Literally a fire truck on water. A boat fully equipped as a fire-fighting vehicle to battle blazes far from shore.

fire buff: A collector, historian or hobbyist with a deep interest and participation in fire apparatus, fire departments and fire fighters.

fire engine: A term loosely used to describe most fire apparatus, but more correctly used to describe a fire vehicle equipped to pump water. Fire department divisions which house only pumpers are called engine companies.

fire fighter: A professional, full-time employed fireman (or woman).

fire truck: A term loosely used to describe most fire apparatus. More correctly used to describe apparatus the primary function of which is to carry ladders, chemicals or personnel. Fire department divisions which do not house pumpers are referred to by their main function; i.e., Ladder Company #5.

fire pole: The familiar brass pole used in fire departments for firemen to travel quickly from the second or third floor living quarters to the garage area during an alarm.

flotation tires: Large extra-wide tires for trucks going into soft terrain to provide added traction and protection from getting stuck.

foam truck: A fire truck with the main duty of carrying and applying foam chemicals to a fire requiring the removal of oxygen supply.

four-wheel drive (4x4) and (6x4): A truck equipped with driving axles in both the front and rear to drive all wheels. Most common on brush trucks, mini-pumpers, air crash trucks and rural areas, especially those with severe weather conditions.

forward-mount pump: A truck in which the pump is mounted in the grill or on a platform on the front bumper. Usually only the smaller commercial chassis.

front-wheel drive: A truck which receives its motive power via the front wheels alone. Many early horse-drawn units were updated in this manner.

gold leaf: A fine application of thinly beaten gold to decorate and stripe fire trucks.

g.p.m.: Gallons per minute. The rating of a pumper's capacity to move water.

ground ladders: Ladders used to reach low windows in fire rescue. Usually carried on the side of most fire trucks.

hand crank siren: Manually operated siren mounted on the cowl or seat of early fire apparatus.

hard rubber tires: Solid tires that require no inflation.

hard suction hose: Stiff, large diameter hose usually connecting the pumper's intake to the hydrant or other water source. Will not collapse under pressure.

hook-and-ladder truck: (see aerial ladder truck) Hook-and-ladder trucks are articulated versions of long ladder-carrying trucks. They sometimes had an additional driver at the rear tiller wheel to help negotiate sharp corners in the city.

hose bed: The area in the fire truck that carries the complement of hoses required for the truck to perform its principal function. Usually directly behind the pump area if the truck is a combination pumper/hose truck.

hose wagon: A fire truck with the main function of carrying hoses.

hybrid: A fire truck that is made up of parts from several manufacturers. Most commercial chassis trucks fall into this category; i.e., Ford/Pierce, Chevrolet/ALF, Reo/Anderson, etc.

insurance patrol: In early days, most fire departments sent out a separate truck and crew to hold water damage down and to help investigate fire causes. They also kept an eye on reducing fire insurance claims and rates.

ladder truck: A fire apparatus with the main function of carrying auxiliary ladders to the scene of a fire.

lantern: Before the advent of electrical components, liquid-fueled lanterns were carried on the back of fire trucks in lieu of taillights. Also useful in search work at night. Highly-prized collector items today.

light truck: A fire truck with the main function of carrying high intensity search lights for night fire fighting.

logs: Record books kept both for fire departments and individual fire engines. The station log books record fire alarms answered, equipment, personnel responding, etc. Each truck in the station also carries its own log book indicating each call answered and every maintenance procedure.

mars light: A red lamp on fire trucks that rotates from side to side, rather than flashing as an emergency signal.

midship pumper: A pumper truck whose pumper is located under the driver's seat or directly behind the cab, ahead of the rear wheels. The most common type of pump placement.

mini-pumper: A small pumper used for small fires. Similar to the brush fighter, although usually thought of as an urban vehicle. Employed frequently by small towns on a budget, though an excellent truck for close-in and fire fighting where maneuverability is desirable.

moto-meter: A temperature gauge mounted in the radiator cap of early motor vehicles. Those on fire trucks are often embellished with the logo of the truck manufacturer.

motor tractor: A powered unit of two or four wheels used to pull a ladder truck trailer or steam pumper unit.

multiple alarm: A fire for which units from several different fire companies must be summoned. The number of companies called will determine whether the fire is a two-alarm, three-alarm, etc.

muster: An event in which fire fighters and fire buffs compete in games of skill with fire apparatus and in static display of fire vehicles.

non-directional tires: A tire which has a very heavy horizontal tread pattern which is angled neither to or from the rear. This type of tire is most commonly found on military vehicles.

out-for-bid: When a fire engine is being sold on the basis of sealed, or open, bids. All parties are invited to submit bids until the closing date when the highest bidder is awarded the truck.

piston pump: A water pump on a fire engine that is operated by pistons rather than rotary gears. The most famous application is the front-mounted Ahrens Fox pumper.

plating: The nickel, chrome or brass covering on most trim or coupling surfaces.

pneumatic tires: tires requiring inflation with air.

pompier ladders: Single-sided ladders used in confined areas where the open end feature is of value. Also allows such ladders to be passed into tighter areas than regular ladders.

pumper: A fire engine with the main function of pumping water through hoses to the fire.

quad: A combination truck that has four major functions; i.e., pumper/ladder/hose/chemical truck.

rescue truck: See emergency truck.

rotary gear pump: A water pump that drives the water through it by the meshing of rotary gears.

scrollwork: The fancy filligree gold leaf decorations on fire engines.

search light: A very large light mounted on the cowl or directly behind the cab of a fire engine for night work. See also light truck.

sedan cab: A fire truck that has the carrying capacity for more than just the front seat passengers. Many have four doors like a family sedan, some have a fully enclosed body like a large panel truck with windows. The city of Detroit was well known for its use of Seagrave sedan cab pumpers for many years.

shaft drive: A truck that is driven between the transmission and differential by a drive shaft instead of a chain or other means.

siren: A loud whining horn operated by hand or electricity to attract attention to an emergency vehicle.

snorkel: A deluge gun or water cannon mounted on the end of a large platform which can be elevated to great heights or over a fire, allowing a fire fighter a greater vantage point when closing in on the fire with high pressure water.

special unit: A truck designed by the fire department to do a special job, such as carrying exhaust fans and ducting.

squad: A fire truck with the main function of carrying fire fighters to the scene of a fire.

station house: The main base for a fire company, often 24-hour living quarters.

steamer: A steam-powered fire pumper. Originally horse-drawn, many such pumpers were later converted to be pulled by a motor tractor.

T-head: A cylinder head design of early truck motors in which exhaust valves were on one side and intake valves on the opposite side.

tiller: The rear steerable wheels of a hook-and-ladder truck.

triple combination: A truck that performs three main functions; i.e., pumper/ladder/hose truck.

triple ignition: A truck whose engine has three sets of spark plugs and three systems of ignition for a fail-safe system.

turntable: The base for an aerial ladder that can be rotated to reach the correct altitude to battle a fire.

turret: See deluge gun.

water cannon: See deluge gun.

water tower: Similar to a snorkel, however the water tower is manned from below, rather than at the gun itself.

volunteer fire company: A fire department which does not have full-time fire fighters. Local residents make up the fire department and respond to emergencies.

Appendix A

Current Manufacturers

The following is a listing of many of the current manufacturers of fire engines in the United States today. Some of them will be of great help to you in the research and restoration of your fire truck, while some couldn't care less about what you are doing as they are strictly interested in selling new fire engines. Most, however, understand that the history you are preserving continues to build confidence and a reputation for dependability for their product. Only manufacturers specializing in fire engines are listed. Manufacturers such as Ford, Chevrolet/GMC, Dodge, Jeep, International Harvester, Kenworth, Peterbilt and others who only supply chassis or other major components to fire engine builders are not represented in this list.

Allegheny Fire Equipment Co.
Huntington, WV

Boardman Co.
Oklahoma City, OK

Brumbaugh Body Co.
Altoona, PA

Calavar Firebird Corp.
Santa Fe Springs, CA

Car-Mar Inc.
Berwick, PA

Crown Coach Corp.
Los Angeles, CA
W.S. Darley & Co.
Melrose Park, IL
Duplex Division of Warner Swasey Co.
Lansing, MI
Farrar Co.
Woodville, MA
Fire Trucks Inc.
Mount Clements, MI
Four Wheel Drive Auto Co.
Clintonville, WI
Forstner Brothers
Medalia, MN
General Safety Equipment Corp.
North Branch, MN
Gerstenslager Corp.
Wooster, OH
Hahn Motors Inc.
Hamberg, PA
Howe Fire Apparatus Co.
Anderson, IN
Imperial Fire Apparatus Co.
Rancocas, NJ
John Bean Division, FMC Corp.
Lansing, MI
Joyce Fire Apparatus
Romulus, MI
Ladder Towers Inc.
Lancaster, PA
Mack Trucks, Inc.
Allentown, PA
Marion Body Works
Marion, WI
Maxim Motor Company
Middleboro, MA
Mobile Aerial Towers, Inc.
Fort Wayne, IN

Peter Pirsch & Sons, Inc.
Kenosha, WI
Pierce Manufacturing Inc.
Appleton, WI
Providence Body Company
Providence, RI
Sanford Fire Apparatus
East Syracuse, NY
Seagrave Fire Apparatus Division, FWD Corp.
Clintonville, WI
Snorkel Fire Equipment Co.
Saint Joseph, MN
Sutphen Fire Equipment Co.
Columbus, OH
Swab Wagon Co.
Elizabethville, PA
Timpco-Seagrave Division, Timmons Metal Products Co.
Columbus, OH
Towers Fire Apparatus Co., Inc.
Freeburg, IL
Truck Cab Manufacturers, Inc.
Cincinnati, OH
United Fire Apparatus Co.
Cridersville, OH
Van Pelt, Inc.
Oakdale, CA
Ward La France Truck Corp.
Long Island City, NY
Waterous Inc.
St. Paul, MN
Welch Fire Equipment Co.
Marion, WI
Yankee-Walter Corp.
Los Angeles, CA
Young Fire Equipment Corp.
Lancaster, NY

Appendix B

Suppliers

Several parts houses are listed to help you in your search for restoration supplies if you can't find what you need locally.

Tools

For tools, any major name brand sold through an auto supply house will serve your purposes. Craftsman (sold through Sears stores), Snap-on and S-K are just a few of the better known names to look for. All offer excellent warranties. Sears in particular still has a no-questions-asked 100% broken tool replacement program. Remember you will be dealing with heavy, sometimes stubborn parts, so a cheap tool may be a broken tool. It can also cost you a broken finger.

Parts

For mechanical parts (Figs. B-1 through B-4) and fire equipment on more recent trucks, contact the manufacturer first: You may be surprised at what is still available directly from the factory. Also contact some of the custom fire truck builders or rebuilders as a source of good used parts. Truck salvage yards and equipment manufacturers often specialize in the conversion of older power plants to modern diesel V8 and other updating of early equipment for fire departments.

You can also check hobby and professional publications listed in Appendix E.

Your local auto parts house can be a great source for parts if you tell them what you are looking for. For example, several years ago I needed a stop light switch for my 1919 American La France pumper. I took the part off the truck and took it to the local auto parts house and asked the counterman for a match of the part. He immediately went to his parts bins and came back with a switch from an early fifties Chevrolet

Fig. B-1. Many fire engine equipment parts, like on this Studebaker, have not changed in years. Checking with a nearby fire equipment parts dealer may produce the needed pieces.

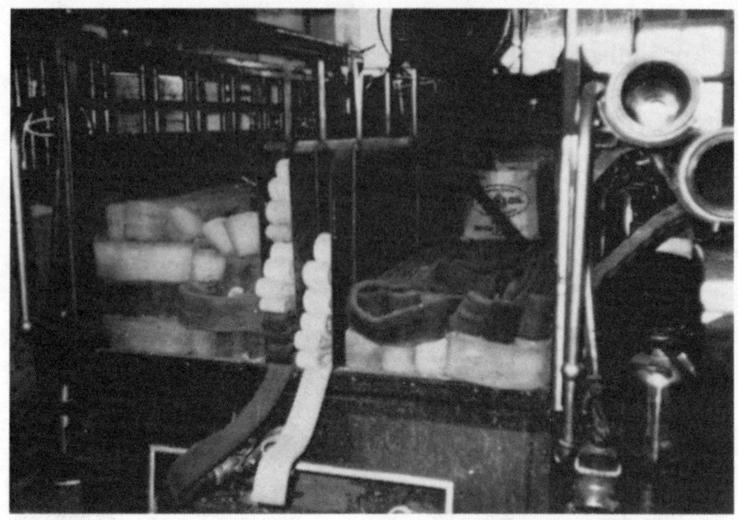

Fig. B-2. Hose and suction pipe can be found at swap meets or perhaps from a fire department where they have been obsoleted. Be careful of rotten hose, however, because a loose hose that breaks under pressure at a muster can cause injuries.

truck that was a perfect match. If I had originally told him what I was really looking for, I am sure he would have thrown up his hands and refused to even look for the part, assuming that no such part existed in his place of business. It is not uncommon for some fire engine parts to be interchangeable with parts of different vehicles.

The following parts houses might be of help if you can't find what you need locally:

Classic Batteries — *batteries*
P.O. Box 13400
Plymouth Meeting, PA 19462
(215) 688-3998

Guy Close, Jr. — *trim clips, fasteners, special screws, etc.*
13426 Valna Dr.
Whittier, CA 90602
(213) 696-3307

The Eastwood Co. *body and fender supplies, tools*
720 E. Lancaster Ave.
Berwyn, PA 19312
(800) 228-2024

Egge Machine Co. *pistons and mechanical parts*
8403 Allport
Santa Fe Springs, CA 90670
(213) 945-3419

Dave Ficken *ignition parts*
P.O. Box 11
Babylon, NY 11702
(516) 587-3332

The Greenland Co. *motor meters, emblems, horns*
P.O. Box 332
Verdugo City, CA 91046
(213) 249-8344

Fig. B-3. Lights, bells and sirens frequently show up at swap meets.

Fig. B-4. This cut-up rear half of a fire engine could produce many needed parts for the careful-eyed restorer.

Gerald Lettieri *gaskets*
132 Old Main St.
Rocky Hill, CT 06087
(203) 520-7177

Maupin Auto Salvage *used parts*
P.O. Box 463
Hutchinson, KS 76501
(316) 665-7951

Metro Molded Parts *rubber parts*
3031 2nd St. N.
Minneapolis, MN 55411

Pierce Manufacturing Co., Inc. *used engines and other parts from fire engine updating*
Doug Ogilvie
P.O. Box 21017
Appleton, WI 54913
(414) 731-5251

Smith & Ryan *Mack parts*
P.O. Box 8
Guild, NH 03754
(603) 863-4107 evenings

Ken Soderbeck *general parts of all types for trucks and equipment*
749 E. South St.
Jackson, MI 49203
(517) 789-6290

Paint and Body Materials

Your local auto body supply company will be your best bet for paint and related materials. You'll find a good selection of primers, surface sealers, paints, sandpapers, masking tapes and even color charts to help you with your work. Companies like 3M which are major suppliers in this business often provide handy how-to pamphlets to help you along. Several of the major paint manufacturers have been in business since the earliest days of spray painting and can offer original paint codes for even the earliest fire trucks. Many paint colors can be color matched in the more modern paint stores. Also, many older fire engine colors have been traditionally one of two shades of red and are even numbered as such; i.e. Red #7, etc. They go back for many years yet are still used on modern trucks in the more traditional departments. Beyond your local auto body supply house, the following supplier can get you most body supplies:

The Eastwood Co. *body supplies*
720 E. Lancaster Ave.
Berwyn, PA 19312
(800) 228-2024

Striping and Letter Supplies

As mentioned earlier, most striping, lettering and gold leaf supplies can be obtained through any good art supply store in your area. Those stores which cater to sign painters will be most helpful. If there is nothing available locally, these companies can help you via mail order:

Dick Blick
P.O. Box 1267
Galesburg, IL 61401
(800) 477-8192
or in Illinois
(800) 322-8183

Buegler Stripers
P.O. Box 29068
Los Angeles, CA 90029

M. Swift & Sons, Inc.
10 Love Lane
Hartford, CT 06112
(203) 522-1181

Tires

Tires for older fire trucks are in short supply. There are currently very few dealers serving this segment of the hobby, as you will see from the following list.

Lucas Automotive, Inc.
P.O. Box 588
Culver City, CA 90230
(213) 397-3732

Coker Tire Co.
P.O. Box 8056
5100 Brainard Rd.
Chattanooga TN 37411
(800) 251-6836 or (615) 892-6018

Bob Calimer
30 E. North St.
Waynesboro, PA 17268
(717) 762-5056

specialist in hard rubber tires

Plating

Most major cities have plating shops, however, it is difficult to find a plater you can trust to treat your parts as you would and to find one who will bother with a fussy restorer and the low volume of your parts. The following companies have proven to be satisfactory for many customers recently.

Christenson Plating Works
2455 E. 3rd St.
Los Angeles, CA 90058
(213) 585-8730

Graves Plating Co.
P.O. Box 1052, Industrial Park
Florence, AL 35630
(205) 764-9487

Master Plating Co.
2109 Newton Ave.
San Diego, CA 92113
(714) 232-3092

Qual Krom
28 Orchard Place
Poughkeepsie, NY 12601
(914) 473-4410

Appendix C

Restoration Shops and Services

The following shops are recommended for complete ground-up restorations of all degrees. Some are specialists who handle only one type or part of the vehicle:

Antique Autos of America
611 Commerce Dr.
Largo, FL 33540
(818) 586-2822
complete and partial restoration

Antique Fire House
P.O. Box 685, 3329 Garfield
Loveland, CO 80537
(303) 667-7040
complete and partial restoration; buy, sell, trade

California Metal Shaping
1704 Hoope Ave.
Los Angeles, CA 90021
(213) 749-5542
custom shaping of sheet metal parts

Cer-Am Glaze
966 86th Ave.
Oakland, CA 94621
(415) 635-2188
porcelain restoration

Classic Coachworks — *complete and partial restoration*
P.O. Box 71, Depot St.
Iola, WI 54945
(715) 445-3648

Drake's Engine Shop — *custom engine rebuilding*
15 Evelyn St.
Rochester, NY 14606
(716) 458-0217

The Generation Gap — *complete and partial restoration*
6148 Hornkensville Rd.
Macon, GA 31206
(912) 788-1111

Tony Nancy — *upholstery*
4363 Woodman Ave.
Sherman Oaks, CA 91423
(213) 789-6923

The Nicolles Wheel Shop — *wooden wheel rebuilders*
P.O. Box 3005
Springfield, IL 62708
(217) 544-9084

Harry Pulfer — *restore, re-enamel and cloissonne emblems*
2700 Mary St.
La Crescenta, CA 91214
(213) 249-3555

Reynolds Head and Block Repair — *casting, welding and repair*
2318 Charles Page Rd.
Tulsa, OK 74127

Schaeffer & Long Inc. — *complete and partial restoration*
210 Davis Rd.
Magnolia, NJ 08049
(609) 784-4044

Ken Soderbeck
749 E. South St.
Jackson, MI 49203
(517) 789-6920

complete and partial restoration, paint and decoration

Joseph Terra
17093 N. Tretheway
Lodi, CA 95240
(209) 369-1386

wooden wheel repair

Vetco of New England
P.O. Box 123
Northfield, MA 01360
(413) 498-2627

complete and partial restoration

Vintage Woodworks
P.O. Box 49
Iola, WI 54945
(715) 445-3791

wood repair and upholstery

White Post Restorations
White Post, VA 22663
(703) 837-1140

complete and partial restorations

Transportation

Transporting a highly prized and/or valuable antique fire engine, whether restored or unrestored, can create unique problems. Weight and size dictate special equipment and the following companies have that equipment and the licensing to move a fire engine for you:

AMR Co.
P.O. Box 824
Kerney, NB 68847
(308) 234-1939

Antique Fire House
P.O. Box 685, 3329 Garfield
Loveland, CO 80537
(303) 667-7040

Horseless Carriage Carriers
61 Iowa Ave.
Paterson, NJ 07503
(800) 631-7796

Passport Transport Ltd.
9479 Aerospace Dr.
St. Louis, MO 63134
(314) 426-6777

Insurance

Don't forget that in most states you will have to insure your antique fire engine if you are going to operate it on the public roads. A stated value policy is usually available from your local insurance agent. State Farm, Allstate and Nationwide are among those who offer such policies. These are not cheap and you will most likely have to have your fire engine appraised before such a policy will be written.

Better possibilities for most fire buffs are the numerous collector car insurance companies. These firms are specialists in unusual vehicles and since they have more experience with the actual risks of insuring these pleasure vehicles, they can usually offer much lower rates. It is best to contact the company before sending in your application and check because some coverages may change without notice and rates fluctuate considerably between companies. Some are also more reluctant than others to insure fire engines.

In any case, be sure to check to see if your policy will cover passengers in your truck. You would not want a passenger to fall off or bump his head on a piece of equipment and end up owning everything you have through a law suit.

Several of the collector vehicle insurance companies are listed.

J.C. Taylor
8701 West Chester Pike
Upper Darby, PA 19082
(215) 528-6450

American Collectors Insurance
P.O. Box 15465
Philadelphia, PA 19149
(609) 234-2552

Condon & Skelly Insurance
P.O. Drawer A
Willingboro, NJ 08046
(609) 871-1212

James Grundy Insurance
500 Office Center D
Fort Washington, PA 19034

Museums

Several of the more prominent museums dedicated to fire apparatus alone or combined with autos of the past are:

Connecticut Fire Museum
P.O. Box 297
Warehouse Pt., CT 06088

Crawford Auto-Aviation Museum
10825 East Blvd.
Cleveland, OH 44106

Edaville Museum
Rt. 58
S. Carver, MA 02366

Fire Museum of Maryland
1301 York Rd.
Lutherville, MD 21093

Four Wheel Drive Museum
Clintonville, WI 54929

Hall of Flame
6101 E. Van Buren Dr.
Phoenix, AZ 85008

Harrah's Auto Collection
P.O. Box 10
Reno, NV 89504

Long Island Auto Museum
Rt. 27, Museum Square
Southampton, NY 11968

Henry Ford Museum
Greenfield Village
Dearborn, MI

Models, Toys and Fire Memorabilia

Automotive Collectables
1366 Lyell Ave.
Rochester, NY 14606
(716) 647-3259

Liberty Art Works *reproduction helmets*
3922 Dover Place
St. Louis, MO 63116

Sinclairs Auto Miniatures
P.O. Box 8410
Erie, PA 16505

Miniature Toys, Inc.
P.O. Drawer E
Westboro, MA 01581

Somes Uniforms
65 Rt. 17
Paramus, NJ 07652
(800) 631-7077
or in New Jersey, (201) 843-1199

Appendix D

Clubs and Organizations

In addition to the comradery of belonging to a fire buffs' organization, the potential for assistance and restoration tips from fellows who have been through it before, the exchange of parts and ideas and the club publications are all of such value as to make memberships in at least a few of these clubs nearly automatic for the new hobbyist.

Antique Automobile Club of America — *$8.50 year, has largest swap meet in the country each October*
501 W. Governor Rd.
Hershey, PA 17033
(717) 534-1910

Antique Truck Club of America — *$8 year*
8-9 115th St.
College Point, NY 11356
(212) 461-1609

Antique Truck Historical Society — *$15 year*
22234 Ford Rd.
Dearborn Heights, MI 48127
(313) 278-7500

Classic Coterie, American La France, Inc. Elmyra, NY 14904	*free to all ALF owners*
Early Ford V-8 Club P.O. Box 2122 San Leondro, CA 94577	*$15 year, Ford cars and trucks 1932-1948*
International Truck Restorers Association 4933 Tacoma Ave. Ft. Wayne, IN 46807 (219) 744-1695	*$10 year, International Harvester trucks of all years.*
Model A Ford Club of America P.O. Box 1791 Whittier, CA 90603	*$8 year, Ford cars and trucks 1928-1931*
Model T Ford Club of America P.O. Box 1711 Oceanside, CA 92054	*$9.50 year, Ford cars and trucks 1909-1927*
Professional Car Society P.O. Box 1976 Garrison, MD 21055 (301) 362-0471	*$10 year, funeral cars, ambulances, emergency vehicles*
The Reo Club of America RD #2 Box 190 Forest, OH 45843 (419) 273-2927	*Reo cars and trucks pre-1936*
Society of Automotive Historians 8201 Woodward Ave. Detroit, MI 48202	*historians of auto and truck industry*

Society for the Preservation and Appreciation of Antique Motor Fire Apparatus in America
P.O. Box 450 Eastwood Sta.
Syracuse, NY 13206
(216) 875-2323

foremost organization on subject of fire engine collecting.

Appendix E

Literature, Periodicals and Books

There are many pieces of literature, books and periodicals available to assist you with your restoration project.

Literature

One vital source of information during your fire engine restoration will be original factory manuals, photos, sales brochures and parts books. Some of these can still be obtained from the original factories, however this is rare. You will usually have to contact specialists who deal in such materials.

Nat Adlestine,
102 Farnsworth
Bordenstown, NJ 08505
(609) 888-1000

Howard & Shelby Applegate
1410 Stallion Lane
West Chester, PA 19380
(215) 692-8034

Firetrucks Unlimited
9959 E. Peakview Ave. No. S-101
Englewood, CO 80111

over 500 fire engine photos in stock catalog $1

Howard & Gladys Hoelschler
4 Pleasant Terrace
Boonton, NJ 07005
(201) 334-8510

Sam Shields
3140 Bayley Rd.
Cuyahoga Falls, OH 44221
(216) 928-9425

Tom Bonsall
P.O. Box 7298
Arlington, VA 22207

Automotive Collectables
1366 Lyell Ave.
Rochester, NY 14606
(716) 647-3259

Periodicals

There are many periodicals that will be of assistance to you both in finding an antique fire truck and in the restoration of it. These products range from hobby weeklies with many ads to annual editions with many permanent listings of suppliers. Most fire buff clubs and organizations also provide periodicals as part of their membership.

Fire Command — *monthly trade journal*, *$10 year*
470 Atlantic
Boston, MA 02210

Firehouse — *monthly trade journal*, *$15 year*
515 Madison Ave.
New York, NY 10022

Hemmings Motor News — *monthly hobby ad paper*, *$9.75 year*
P.O. Box 100
Bennington, VT 05201

Old Cars Weekly — *weekly hobby newspaper*
700 E. State St. — *$13.50 year*
Iola, WI 54945

Skinned Knuckles — *monthly restorers' magazine*
175 May Ave. — *$6 year*
Monrovia, CA 91016

Vintage Auto Almanac — *annual, $4.95*
P.O. Box 945
Bennington, VT 05201

The Visiting Fireman — *annual hobby directory*
203 N. Washington Ave. — *$5.50*
Batavia, IL 60510

Western Fire Journal — *monthly trade journal*
9072 E. Artesia Blvd., Suite #7 — *$8 year*
Bellflower, CA 90706

Wisconsin Fire Journal — *monthly trade journal*
4229 Beltline Highway — *$6 year*
Madison, WI 53711

Books

A few books for suggested reading on fire engines in general and restoration of antique fire trucks are listed.

Car Interior Restorations — $3.95, TAB BOOKS Inc.
by Terry Boyce — Blue Ridge Summit
PA 17214

101 Tips and Tricks for Car Restorers — $3.95, TAB BOOKS Inc.
by William Cannon — Blue Ridge Summit
and Ron Bishop — PA 17214

Collecting and Building Model Trucks
by Lou Kroack
$9.95, TAB BOOKS Inc.
Blue Ridge Summit
PA 17214

Chilton's Auto Restoration Guide
by Burt Mills
$15, Chilton Book Co.
Radnor, PA 19087

The Complete Encyclopedia of Commercial Vehicles
by G.N. Georgano
$29.95, Krause Pubs.
700 E. State St.
Iola, WI 54945

American Fire Engines Since 1900
by Walt McCall
$24.95, Classic Motorbooks
P.O. Box 1F1B
Osceola, WI 54020

Seagrave Fire Apparatus 1900-1930
$9.95, Classic Motorbooks
P.O. Box 1F1B
Osceola, WI 54020

Index